SpringerBriefs in Physics

SpringerBriefs in Physics are a series of slim high-quality publications encompassing the entire spectrum of physics. Manuscripts for SpringerBriefs in Physics will be evaluated by Springer and by members of the Editorial Board. Proposals and other communication should be sent to your Publishing Editors at Springer.

Featuring compact volumes of 50 to 125 pages (approximately 20,000–45,000 words), Briefs are shorter than a conventional book but longer than a journal article. Thus, Briefs serve as timely, concise tools for students, researchers, and professionals.

Typical texts for publication might include:

- A snapshot review of the current state of a hot or emerging field
- A concise introduction to core concepts that students must understand in order to make independent contributions
- An extended research report giving more details and discussion than is possible in a conventional journal article
- A manual describing underlying principles and best practices for an experimental technique
- An essay exploring new ideas within physics, related philosophical issues, or broader topics such as science and society

Briefs allow authors to present their ideas and readers to absorb them with minimal time investment. Briefs will be published as part of Springer's eBook collection, with millions of users worldwide. In addition, they will be available, just like other books, for individual print and electronic purchase. Briefs are characterized by fast, global electronic dissemination, straightforward publishing agreements, easy-to-use manuscript preparation and formatting guidelines, and expedited production schedules. We aim for publication 8–12 weeks after acceptance.

More information about this series at http://www.springer.com/series/8902

Saranya Samik Ghosh · Thomas Hebbeker ·
Arnd Meyer · Tobias Pook

General Model Independent Searches for Physics Beyond the Standard Model

 Springer

Saranya Samik Ghosh
Physikalisches Institut III A
RWTH Aachen University
Aachen, Nordrhein-Westfalen, Germany

Thomas Hebbeker
Physikalisches Institut III A
RWTH Aachen University
Aachen, Nordrhein-Westfalen, Germany

Arnd Meyer
Physikalisches Institut III A
RWTH Aachen University
Aachen, Nordrhein-Westfalen, Germany

Tobias Pook
Physikalisches Institut III A
RWTH Aachen University
Aachen, Nordrhein-Westfalen, Germany

ISSN 2191-5423   ISSN 2191-5431   (electronic)
SpringerBriefs in Physics
ISBN 978-3-030-53782-1   ISBN 978-3-030-53783-8   (eBook)
https://doi.org/10.1007/978-3-030-53783-8

This Springer imprint is published by the registered company Springer Nature Switzerland AG
The registered company address is: Gewerbestrasse 11, 6330 Cham, Switzerland

Preface

Since the 1960s, there have been several compelling theoretical predictions for new fundamental particles, for example, the gluon or the top quark, which were confirmed by experiment. These observations were achieved with the help of particle colliders with higher and even higher energies. Most recently the Higgs boson was discovered by the ATLAS and CMS experiments at the Large Hadron Collider (LHC) at CERN.

The Standard Model of particle physics (SM) has been developed over several decades, and it is now highly successful in describing the fundamental particles and the interactions between them. It has passed a multitude of experimental tests over the past 40 years or so. However, we know that the SM is at best an incomplete description of the universe, since there are several phenomena that remain unexplained, for example, the matter-antimatter asymmetry in the universe.

Several models of new physics beyond the Standard Model (BSM) have been proposed to address the shortcomings of the SM. Examples are Supersymmetry and extra-dimensional models. So far, no compelling experimental evidence has been found to favour any of the popular BSM physics models, and large regions of the parameter space of the proposed models have been experimentally excluded. Yet, the search for a better understanding of the fundamental particles and their interactions carries on. The searches for new physics performed at collider experiments are typically developed around specific BSM model predictions, with the investigated final states and kinematic selections being chosen and optimised to have sensitivity to a particular model or a limited number of models. Although a large number of dedicated analyses corresponding to many new physics models have been and are being performed, there still remain numerous different models and parameter space regions within the considered models that are unexplored due to limitations on experimental resources and person-power.

We are now at a stage where there is no particular compelling specific theoretical prediction to go after. On the other hand, there are still vast amounts of unexplored data that have been and are expected to be collected at current and future experiments that may contain the elusive evidence for something new. Therefore, physicists must employ novel data analysis methods to extract the most out of these large datasets.

The aforementioned limitations on **dedicated search analyses** highlight the importance of developing **general model independent approaches** to search for new physics in the large datasets available for analysis. Furthermore, such model independent searches are potentially sensitive to physics phenomena that have not been included in any of the currently theorised models and hence are likely to be overlooked by model-driven searches. The model independent searches we consider here investigate a large dataset, ideally the full data set recorded by a given experiment. Their search covers a large range of final states and kinematic regimes, without optimising for a dedicated BSM scenario. Anomaly detection and investigation form a key part of such an approach.

In this review, we present and compare the past and current general model independent search strategies and their results, and we give an outlook at future developments. While these searches pertain to particle physics, the techniques developed and utilised in these searches are common to searches for new models in other fields as well.

Aachen, Germany   Saranya Samik Ghosh
July 2020   Thomas Hebbeker
                                                              Arnd Meyer
                                                             Tobias Pook

Acknowledgments

We would like to thank the following former master and doctoral students at RWTH Aachen University for their inputs about model independent searches, while contributing to the MUSiC project: Andreas Albert, Philipp Biallass, Michael Brodski, Deborah Duchardt, Carsten Hof, Erik Dietz-Laursonn, Simon Knutzen, Mark Olschewski, Holger Pieta, Jonas Roemer, Stefan Schmitz.

We profited a lot from discussions of our colleagues Sascha Caron and Albert de Roeck, and we want to express our appreciation here.

Contents

Chapter 1
Introduction

Particle physics appears to be entering a new era again. Since the 1960s, it has often been the case that there have been certain compelling theoretical predictions, such as predictions for new fundamental particles, and particle colliders with higher and even higher energies were constructed leading to the discoveries of the predicted particles. This has been the case over the years leading to, most recently, the discovery of the Higgs boson by the ATLAS and CMS experiments at the Large Hadron Collider (LHC). However, we are now at a stage when there is no particular compelling theoretical prediction to go after, with the lack of any major hints of new physics phenomena following the discovery of the Higgs boson. Fortunately, there are still vast amounts of unexplored data that have been and are expected to be collected at current and future experiments that may contain the elusive evidence for something new. The unprecedented amount of experimental data expected to be collected at the High-Luminosity LHC being a prime example of this. Physicists must employ novel data analysis techniques to extract the most out of these large datasets, and, of late, there has been an increased interest in automated statistical analysis of data not just in physics but all over academia. Model independent searches for new physics build upon the data analysis tools available to search for new undiscovered phenomena that may lie hidden within the experimental data, and the motivation for such searches is becoming increasingly more compelling.

The Standard Model of particle physics (SM) has been highly successful in describing the fundamental particles and the interactions between them, while standing the test of several experimental measurements over the past 40 years or so. Several of the predictions of the SM have been experimentally verified, with the high profile discovery of a particle consistent with the predicted Higgs boson being a recent validation of the SM. However, we know that the SM is at best an incomplete description of the universe since there are several phenomena that remain unexplained by the SM. The lack of understanding of the matter-antimatter asymmetry in the universe, the unexplained masses of neutrinos, the lack of an explanation for dark matter and

© The Author(s), under exclusive license to Springer Nature Switzerland AG 2020
S. S. Ghosh et al., *General Model Independent Searches for Physics Beyond
the Standard Model*, SpringerBriefs in Physics,
https://doi.org/10.1007/978-3-030-53783-8_1

the inability of the SM to describe gravitational interactions feature among the major shortcomings of the SM.

Several models of new physics beyond the Standard Model (BSM) have been proposed to address the shortcomings of the SM. Supersymmetry, extra-dimensional models, models extending the SM to include neutrino masses, and a wide range of other models are competing to take the place of the SM. So far, no compelling experimental evidence has been found to favour any of the popular BSM physics models, and several regions of the phase space of the proposed models have been experimentally excluded. Yet, the search for a better understanding of the fundamental particles and their interactions carries on.

There are a number of experimental searches dedicated to examining several of the BSM physics models, in dedicated experiments as well as general collider experiments such as those at the LHC. The searches for new physics performed at collider experiments are typically developed around specific BSM model predictions, with the investigated final states and kinematic selections being chosen and optimised to have sensitivity to a particular model or a few limited models. Although a large number of dedicated analyses corresponding to different new physics models have been and are being performed, there still remain numerous different models or phase spaces within the explored models that remain unexplored due to limitations on experimental resources and person-power.

The aforementioned limitations on dedicated search analysis, the sizeable number and variety of proposed BSM physics models, and the large datasets collected by several experiments, in particular by collider experiments including those at the LHC, highlight the importance of developing general model independent approaches to search for new physics. Such searches aim to identify deviations in the data and detect anomalies compared to the SM expectations without specific inputs from any particular models of BSM physics and are thus potentially sensitive to a wide range of signatures associated with a variety of new physics models. Furthermore, model independent searches are potentially sensitive to unforeseen physics phenomena that have not been included in any of the currently theorised models and hence are likely to be overlooked by model-driven search strategies. A model independent search covers a large range of final states and kinematic regimes without optimising for a dedicated BSM scenario. An understanding of the detector performance and the validity of the simulation that describes the SM prediction are vital to such search approaches, since it is often only these external inputs other than the dataset itself that is used to have a robust and unbiased general search strategy.

While the model independent searches pertaining to particle physics are discussed here, the techniques developed and utilised in these searches are common to searches for new models in other fields as well.

This document will first discuss the motivation for a general model independent search for new physics by describing in further detail the current theoretical understanding of particle physics and providing an overview of experimental searches performed endeavouring to find new physics in past and present high energy physics experiments. Further, the concepts and principles behind the general model independent search for new physics, particularly at collider experiments will be described.

Then, past instances of model independent searches that were performed will be described followed by a discussion of the current status of model independent searches, particularly in the context of the experiments at the LHC. The following chapter compares and evaluates those methods and results. Finally, the future prospects of such general searches, with possible improvements using new tools such as machine learning techniques, will be discussed.

Each section describing a certain method or a given experiment is written such that it can be read as a 'standalone' text. For kinematic quantities and formulae, we have always kept the notation used by the experiment under discussion. Only in Chap. 6 do we move to a common nomenclature.

Chapter 2
Motivation for General Model Independent Search for New Physics

A brief overview of the current status of the field of high energy physics is described in this chapter, with a very short introduction to the Standard Model (SM) of particle physics, and the motivation that physicists have for hypothesising new physics beyond the Standard Model (BSM) with examples of some of the more popular BSM theories. Further, a short discussion on the current experimental searches for BSM physics is presented, providing the context for the discussion on the motivation for developing a general model independent search strategy for new physics beyond the Standard Model.

2.1 Standard Model of Particle Physics

The Standard Model (SM) of particle physics can be said to be the pinnacle in terms of the theoretical understanding in the field of high energy physics till date. It describes the elementary particles that make up the matter in the universe and the interactions between them, and though it cannot be considered a complete theory, it is the most comprehensive one to have stood up to experimental scrutiny as yet.

The SM is a paradigm of a Quantum Field Theory (QFT), wherein the different particles are associated with their fields. The elementary matter particles described by the SM are grouped into quarks and leptons, with the quarks consisting of the up (u), down (d), strange (s), charm (c), top (t), and bottom (b) quarks that make up the baryons in the universe, whereas the leptons consist of the electron (e), muon (μ), tau (τ) and the corresponding neutrinos $(\nu_e, \nu_\mu, \nu_\tau)$ along with their anti-particles. All of these particles are fermions.

Of the four fundamental forces of nature, only gravity is not described by the SM, with the others being the electromagnetic force that is described in terms of Quantum

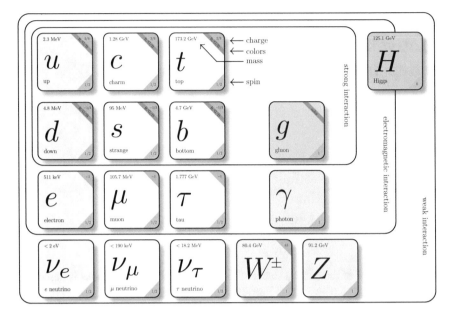

Fig. 2.1 Fundamental particles of the standard model [7]

Electrodynamics (or QED) in the SM, the weak force that is described in terms of quantum flavordynamics (or QFD, also known as electroweak interactions together with QED) in the SM, and the strong force that is described in terms of Quantum Chromodynamics (QCD) in the SM. The description of each of the fundamental interactions described in the SM is in terms of fields, and as such there exist carrier particles associated with each of the fields that are bosons. The photons (γ) are the carriers of QED and thus the electromagnetic interactions, whereas the W^+, W^- and Z^0 bosons are the particles associated with the weak interactions and the gluons (g) are associated with QCD or the strong interactions.

Along with the aforementioned particles and fields there exists another field, the Brout–Englert–Higgs (BEH) field that was proposed in order to explain the mechanism of electroweak symmetry breaking which leads to the origin of masses for the fundamental particles [1–4]. The particle associated with this field is called the Higgs boson (H), and it was the final of the particles predicted by the SM to be experimentally discovered, when it was first observed by the ATLAS and CMS experiments at the Large Hadron Collider (LHC) in 2012 [5, 6]. All the fundamental particles of the Standard Model are shown in Fig. 2.1.

The SM came to resemble its current form closely in the 1970s, and since then it has stood up to significant experimental scrutiny, from the discovery of weak neutral currents at the Gargamelle bubble chamber at CERN in 1973 [8–10] and the discovery of the $W^\pm$ and Z^0 bosons at the UA1 and UA2 experiments [10–13]. Subsequently, the various particles predicted by the SM have been discovered experimentally, with the discovery of the Higgs boson in 2012 completing the final piece of the puzzle.

While the SM provides, at present, our best fundamental understanding of the phenomenology of particle physics, it has several shortcomings because of which it does not completely describe our observed universe. This leads many physicists to believe that the Standard Model is just the low energy limit of a more fundamental theory of nature.

2.2 Shortcomings of the Standard Model

The ultimate goal of particle physicists is to have one unified theory that describes nature with all the particles in the universe and their interactions, sometimes referred to as a Theory Of Everything (TOE). The Standard Model falls short on that front, and some of the major shortcomings are described below.

1. The SM does not describe gravitational interactions. Gravitation is one of the four fundamental forces of nature, that is essential in order to explain the universe at larger scales; however, it is the weakest of the forces and is not as relevant at the level of elementary particles, particularly in comparison with the other forces. While the SM does describe the other forces, gravitational force is not a part of the SM. The general theory of relativity remains to be the most consistent explanation of gravity, and as yet there has not been an adequate description of gravity in terms of a quantum field theory. There exist BSM physics models that hypothesise a graviton particle that would be associated with gravitational interactions; however, no experimental evidence of the existence of such a particle has been found yet.

2. The SM does not have an explanation for dark matter. Cosmological observation and the prevailing cosmological model, the Λ-CDM model, that recent cosmological measurements agree with [14], describe a universe where less than 5% is made up of the observed ordinary matter that the SM describes. Dark matter, particularly Cold Dark Matter (CDM), is believed to constitute around five times the ordinary matter content of the universe. The remaining part of the universe is explained by the cosmological constant Λ that is associated with dark energy. Astronomical observations such as galactic rotation curves and studies of collision of galaxies also point towards the presence of dark matter. The SM does not include any candidate particles that might explain the dark matter or dark energy content of the universe. Certain hypothesised BSM models do include potential dark matter candidate particles; however, there is no experimental evidence supporting these till date.

3. The SM does not account for the mass of neutrinos. The discovery of neutrino oscillations [15, 16], which requires neutrinos to be massive in nature, led to the Nobel Prize in Physics being awarded to the representatives of the SNO and Super-Kamiokande collaborations. According to the SM, neutrinos are supposed

to be massless and any theoretical explanation for neutrino masses would require the SM to be extended from its current form.

4. The SM does not explain the matter-antimatter asymmetry in the universe. It is known that in the observed universe, there is a predominance of matter over antimatter, and the SM does not have an explanation for such a large asymmetry between matter and antimatter. While it has been argued from a theoretical standpoint that the Standard Model should contain a stronger breaking of the Charge-Parity (CP) symmetry that relates matter to antimatter, particularly in the case of the strong interaction. However, no such violation of the CP symmetry has been found experimentally. This is also referred to as the strong CP problem.

5. The SM does not address the hierarchy problem. The experimentally observed mass of the Higgs boson (mass of $125\,\text{GeV}$) is of the same order of magnitude as that of the carriers of the weak interaction, that is the $W^\pm$ bosons (mass of $80\,\text{GeV}$) and Z^0 bosons (mass of $91\,\text{GeV}$), and this scale of the order of $100\,\text{GeV}$ is the weak scale. The Planck scale, which is associated with gravitational interactions, is of the order of $10^{18}\,\text{GeV}$. The vast difference in these scales is referred to as the hierarchy problem, and it is considered a problem because in order for the weak scale to be so much smaller than the Planck scale, severe fine tuning of the parameters of the SM is required. Such fine tuning is not desirable for a robust theory of nature and is seen to violate the principles of naturalness.

The points illustrated above provide the context for the motivation and requirement to explore physics beyond the Standard Model in order to address the shortcomings of the SM and to have a more comprehensive theory of nature.

2.3 Theories of Physics Beyond the Standard Model (BSM)

Having motivated the need to look beyond the SM, it appears reasonable that physicists consider the SM to represent the low energy limit of a more complete theory of nature, be that the Grand Unified Theory (GUT), which would unify fundamental interactions described by the SM into one, or the theory of everything that aims to be a complete description of nature including gravitational interaction. While there are some fundamental theories that have been developed to address such aspirations, most notably the string theory, some of them cannot be probed by current day particle physics experiments. There are several prominent phenomenological models of physics beyond the SM that address some of the shortcomings of the SM. Many of these have been experimentally probed by existing particle physics experiments. Since the SM has been experimentally verified despite its shortcomings, many of the BSM models propose extensions to the SM with additional particles or other departures in the properties not tested with the required precision as yet.

Models based on supersymmetry (SUSY) have been seen to be the most prominent candidates for BSM physics. Supersymmetry introduces a symmetry between the two different classes of particles that are fermions (particles with half of a unit of spin or odd multiples of it) and bosons (particles with integral units of spin). In the SM, the matter particles make up the fermions, whereas the interaction particles make up the bosons. Supersymmetry predicts that each of the particles in the Standard Model has a partner, referred to as the super-partner, with a spin that differs by half of a unit and is related to the original particle by a supersymmetry transformation. Supersymmetry must be a broken symmetry, because exact supersymmetry requires every super-partner to be degenerate in mass with its corresponding SM particle, and if that were the case then these super-partner particles would have been discovered by the experimental searches that have already been conducted. A more detailed overview of supersymmetry can be found in Ref. [17]. Several BSM models are based on supersymmetry, such as the MSSM (Minimal Supersymmetric Standard Model) that proposes only the most basic extensions to the SM required to include supersymmetry. There are also several models based on further extending the SM based on supersymmetric principles. Such models aim to fulfil several of the shortcomings of the SM, including addressing the hierarchy problem and certain models predict the lightest supersymmetric particle to be stable and electrically neutral, interacting weakly with the particles of the SM, thus fulfilling the characteristics required for a dark matter candidate. The supersymmetry framework can also be compatible with other higher energy theories such as grand unification and string theory. Certain supersymmetric models also include gravitational interactions, in the SuperGravity or SUGRA framework, with the mSUGRA (minimal SUGRA) models using the minimal extensions to include supergravity [18–20]. To account for the absence of evidence for supersymmetry, models predict that the super-partner particles must have a mass higher than the experimental reach of past experiments. Models based on low energy supersymmetry predict the existence of these super-partner particles with a mass in the TeV range, which could potentially be probed, and such particles might be discovered at current experiments such as those at the Large Hadron Collider (LHC). Since there are several different models based on the supersymmetry, these can often have different signatures which complicate the efforts in terms of experimental searches.

Further, there are the BSM models that provide an explanation for neutrino masses, which can be based on the seesaw mechanism [20–22]. The seesaw mechanism provides an attractive explanation of the neutrino masses and their smallness when compared to the masses of the charged fermions of the same generation through the existence of heavy neutrinos that are yet to be discovered. There are multiple variations of this basic premise in different models. The prediction of the existence of heavier neutrino like particles by these models is something that is being probed at several experiments. Based on the properties of the new particles predicted, they could also be dark matter candidates.

Another class of BSM models is those based on the existence of extra dimensions in addition to the three space and one time dimension of the known universe. The idea of extra dimensions was proposed as far back as in the 1920s by Kaluza and Klein [23, 24]; however, recent phenomenological models based on extra dimensions are, e.g. based on the proposals of Arkani-Hammed, Dimopoulos and D'vali (ADD) [25–27]. Explanations for the observed universe showing only the three spatial and one time dimension can be proposed as if our observable world is constrained to exist on a four-dimensional hypersurface (called the brane, or membrane) while being embedded in a higher dimensional space (called the bulk). These BSM models can provide an explanation for some of the shortcomings of the SM such as the hierarchy problem or gravitational interactions, an overview of such models can be found in Ref. [28]. Different variations of such extra-dimensional models come with different signatures that could be detected at experiments, for example, microscopic black holes, gravitons or additional particles in the form of the so-called Kaluza–Klein partner particles of different SM particles.

Other prominent models of BSM physics include those based on the Little Higgs models, where the Higgs boson is taken to be a composite particle [29] and some of these models propose the existence of new particles with masses around the 1 TeV scale that could be subject to experimental scrutiny, or models that propose the existence of new types of particles such as axions, which are hypothetical particles that could address the strong CP problem and provide candidates for cold dark matter [30]. Other particles like leptoquarks might also exist, which are hypothetical particles carrying both baryon number and lepton number and hence could allow quarks and leptons to interact, thus providing grounds for the unification of the electroweak and strong interactions.

There are several other important BSM physics models that have not been listed here for the sake of brevity. It is also important to keep in mind that it is possible that the true form of nature is not adequately modelled by any of the BSM models that have been proposed and new physics phenomena might have different signatures compared to those predicted by the BSM models proposed yet.

The purpose of this section is to highlight the wide landscape of different new physics models that have been proposed and are being experimentally probed. It is indeed a challenging task for experiments to probe the wide scope of BSM physics already out there, and further work is being done on the theoretical side. For each paradigm mentioned above, there are several BSM models with different experimental signatures associated with them, and that is still very far from an exhaustive list. No direct experimental evidence for any of these BSM models has been found till date; however, experimental searches continue to be conducted in the hope of finding hints of new physics phenomena.

2.4 Experimental Searches for BSM Physics

The aim of addressing the shortcomings of the SM has lead to the development of a plethora of BSM models; however, no BSM physics model can be accepted without experimental confirmation. Several different high energy physics experiments have searched for signatures of new physics in the past and several continue to do so at present. Several BSM physics models have been probed by these experiments but no compelling evidence supporting any particular BSM model has been found yet. A few of the prominent particle physics experiments searching for new physics are mentioned below, starting with the non-collider-based experiments, followed by the major collider-based experiments, providing the context for a general model independent search approach that can be implemented at collider-based experiments.

2.4.1 Searches at Non-Collider-Based Experiments

There are several dedicated particle physics experiments that are designed to search for BSM physics in terms of searching for hypothesised new particles, such as axions, or to study a particular class of particles, such as neutrinos, in order to explore new BSM models through them. These experiments tend to be non-collider-based experiments, and some of the experiments that are in operation are discussed here.

Neutrino-based experiments are one set of such dedicated experiments. The discovery of neutrino oscillations and the massive nature of neutrinos [15, 16] by the SNO and Super-Kamiokande experiments posed an important experimental challenge for the SM that does not explain such phenomena. Current day neutrino experiments are studying the phenomena of neutrino oscillations in greater detail, to confirm oscillation between the different combination of flavours, and make precise measurements of the difference in mass between the different neutrinos along with attempting to establish the magnitude of the neutrino masses for the different neutrino flavours. Other major goals of certain neutrino experiments include searches for sterile neutrinos or other dark matter candidate particles and attempting to test whether neutrinos are Dirac or Majorana particles, that is whether they are their own anti-particles or not. Several BSM theories have different predictions or assumptions for the properties under test at these experiments, and hence these experiments form an important test for such models. Borexino [31], Daya Bay [32], Double Chooz [33], KamLAND [34] and T2K [35] are just a few of the major neutrino-based experiments operating now and there are several experiments that are expected to become operational in the near future, such as JUNO [36] and DUNE [37]. IceCube [38] and ANTARES [39] are also prominent neutrino experiments and these experiments have a strong programme in the field of astronomy along with particle physics.

Several experiments dedicated to the search for dark matter candidate particles have been in operation and more are being planned. The general idea of such experiments is to have particle detectors situated underground or in a location where there is very little background in terms of particle flux from cosmic or other radiation sources in order to detect the interactions of possible dark matter candidate particles. Some experiments in or near operation are XENON [40], LUX [41], DAMA [42], CDMS [43, 44], DARWIN [45] in their different incarnations, and this is a non-exhaustive list.

There are also the experiments designed to search for Axions or similar particles, such as CAST [46], ADMX [47], MADMAX [48] and ABRACADABRA [49], with more experiments planned to start operations in the near future.

This section does not cover the breadth of dedicated particle physics experiments searching for BSM physics, it is intended to provide a glance at the large experimental effort that has already been exerted to search for BSM physics at non-collider-based experiments.

2.4.2 Searches at Collider-Based Experiments

Particle accelerators are complex machines that are designed to accelerate particles to very high speeds, close to the speed of light, thus imparting very high momenta and energies to these particles. Two beams of such accelerated particles travelling in opposite directions are made to collide at the so-called interaction points where particle detector-based experiments are placed, thus creating particle collisions with very high center-of-mass energies. Such high energy particle collisions lead to the production of fundamental particles and can possibly produce particles hypothesised by different BSM models, thus making collider-based experiments an ideal setting to search for new physics beyond the Standard Model. A wide variety of final states, consisting of a combination of different final state particles are produced in collider experiments, and processes that are part of the SM constitute backgrounds to the signatures of possible new physics processes. The observable final state particles are mostly decay products of the short-lived particles produced in the collisions. Simulations of SM processes are often used to estimate the background from the SM processes. Since there are a multitude of BSM models, predicting different signatures in a range of different final states, it becomes challenging to search for the signs of BSM physics while dealing with backgrounds in numerous different final states.

The SM was already probed at the Super Proton–Antiproton Synchrotron ($S p \bar{p} S$) at CERN in the 1980s where proton and antiprotons were collided at center-of-mass energy of 315 GeV, and later at close to 900 GeV after upgrades. It was there that UA1 and UA2 experiments were based, that led to the first experimental discovery of the $W^{\pm}$ and Z^0 bosons of the SM [11–13, 50] in 1983. It was slightly before then, in 1979, that the Positron–Electron Tandem Ring Accelerator (PETRA) at DESY in Hamburg, Germany, where electrons and positrons were collided that the first direct experimental evidence for gluons was found by the TASSO, MARK J, PLUTO and JADE experiments [51–54], leading to experimental support to the establishment of QCD as a theory of strong interactions. There were also several important collider experiments before then, such as at the SPEAR collider at SLAC, Stanford.

More recently, the experiments at the Large Electron–Positron Collider (LEP) at CERN made important measurements of SM physics and also had a substantial programme to search for BSM physics signatures. LEP collided electrons and positrons at center-of-mass energies of up to around 200 GeV between 1989 and 2000 with the major particle detector-based experiments associated with LEP being the ALEPH [55], DELPHI [56], L3 [57] and OPAL [58] experiments. The Tevatron collider at Fermi National Accelerator Laboratory, USA was a proton–antiproton collider at a center-of-mass energy of close to 2 TeV that ran between 1983 and 2011 with the major collider experiments of CDF [59] and D0 [60, 61] associated with it. While these experiments are more renowned for the first experimental discovery of the top quark [62, 63], they also had a significant physics programme aimed at searching for BSM physics. Another important particle collider that hosted experiments with a BSM search programme was the Hadron-Electron Ring Accelerator (HERA) at DESY, Hamburg where electrons or positrons were collided with protons at a center-of-mass energy of up to 318 GeV and which operated from 1992 until 2007. The major general experiments at HERA were the H1 [64] and ZEUS experiments. The BaBar experiment [65] at the PEP-II collider at SLAC, USA and the Belle experiment [66] at the KEKB collider at KEK, Japan, were two experiments at high intensity electron–positron colliders running well into the 2000s that were focused mainly on studying B-mesons but also conducting BSM searches, with the latter being upgraded and starting again for a second run.

The Large Hadron Collider (LHC) at CERN is the most powerful particle accelerator in operation today, and it has been designed to collide protons with protons at a center-of-mass energy of 14 TeV, with large datasets of proton–proton collision at center-of-mass energies of 7, 8 and 13 TeV having already been recorded by the LHC-based experiments since the beginning of operations in 2010. The major experiments based at the LHC which have a large physics programme devoted to the search for BSM physics are the A Toroidal LHC ApparatuS (ATLAS) experiment [67] and the Compact Muon Solenoid (CMS) experiment [68], which consist of general-purpose particle detectors, and Large Hadron Collider beauty experiment (LHCb) [69] that primarily focuses on the physics of b-hadrons. While the discovery of the Higgs boson by the ATLAS and CMS experiments [5, 6] has been the highlight of the physics programme at the LHC so far, several diverse analyses searching for BSM physics have been and continue to be conducted by the LHC-based experiments.

There are also proposals to build further colliders in the future, such as the International Linear Collider (ILC) or the Future Circular Collider (FCC), and even though these are not yet finalised, the collider physics programme is sure to continue when the LHC itself will be upgraded after 2024 into the High-Luminosity LHC, in order to deliver several times more data than what has been recorded till then. This underlines the scope for conducting experimental searches for BSM physics phenomena at collider experiments in the future.

2.5 Motivation for General Model Independent Search for New Physics

In the preceding sections, it has been discussed that the Standard Model of particles physics is the established theory that describes the fundamental particles and their interactions. However, the SM has several very important shortcomings, and hence it is not considered to be a complete theory of nature. Hence, it is evident that there must be new physics beyond the Standard Model that is yet to be discovered.

The previous sections also discuss that, in order to address the shortcomings of the SM, physicists have proposed models of physics beyond the Standard Model and there are a multitude of well-motivated BSM models that have been proposed and several of these BSM models have been probed at different particle physics experiments. Despite the substantial experiment programme already in place, no direct evidence favouring any model of BSM physics has turned up yet, and such searches are continuing at the high energy physics experiments that are currently in operation.

Since there are a large number of BSM models with different potential experimental signatures, it is not feasible to conduct experimental searches for each of the proposed models. Hence a model independent search becomes desirable, where the search is not just limited to one or a few BSM models. Also, the true form of BSM physics phenomena might not be adequately described by any of the currently proposed BSM models, and experimental searches focusing on only the signatures of the currently proposed BSM models might miss signs of new physics that could be hidden in the data in regions of the phase space left unexplored. Furthermore, collider-based experiments with a general-purpose particle detector, for instance, the ATLAS or CMS experiments at the LHC, are able to detect several final state particles, and hence access several different final states and potentially probe a wide range of BSM models with varied experimental signatures. While it is not practically feasible to study each final state individually, a general search that probes several different final states with an automated framework, not confined by the scope of a particular BSM model, can be a powerful tool to search for new physics. General model independent searches at collider experiments aim to do just that, i.e. to search for signatures of physics beyond the Standard Model in multiple different final states without having any constraints or specialisation based on any particular model of BSM physics.

A more detailed description of the concept of a general model independent search along with discussions of the benefits and drawbacks of such analyses and descriptions of the implementation of such an approach at some of the collider experiments can be found in the following chapters.

Chapter 3
Concept of General Model Independent Searches for New Physics

Having motivated the need for conducting general model independent searches for new physics at collider-based experiments in the previous chapter, this chapter contains a description of the concept of such an approach along with comparisons to the dedicated search analyses that are conducted to target specific BSM models, discussing both the benefits and drawbacks of the different approaches and then emphasising the complementary nature of the two approaches.

3.1 Description of the General Model Independent Search Approach

In a nutshell, the strategy of the model independent approach is to search for signs of new physics beyond the Standard Model in an unbiased manner without the input of any specific BSM model, relying purely on the predictions of the SM while also taking into account relevant experimental effects. A general search at a particle physics experiment is expected to search in a wide range of final states and phase space regions. In order to be able to do this, given the constraints in terms of human resources and computing power available, a general approach is required that can be automated so that a large number of final states or phase space regions can be explored without requiring human intervention for each region explored. In each final state or phase space region, the search algorithm used will then search for deviations in the experimental data compared to the background expectation. The background expectation is taken based on the predictions of the SM processes. Any deviation in the recorded data compared to the predictions of the SM is then to be identified and examined as a possible sign of new physics. This approach is described in more detail below.

The collider-based general experiments are able to detect final state particles that are stable within the volume of the detector such as electrons, muons, photons and several hadrons or jets of particles that are formed during the hadronization of quarks

© The Author(s), under exclusive license to Springer Nature Switzerland AG 2020
S. S. Ghosh et al., *General Model Independent Searches for Physics Beyond the Standard Model*, SpringerBriefs in Physics,
https://doi.org/10.1007/978-3-030-53783-8_3

and gluons along with missing transverse energy. The missing transverse energy is the net momentum imbalance in the detector transverse plane that is associated with particles that can go undetected in the detector due to very weak interaction as is the case for neutrinos. Several important kinematic observables of these final state particles, such as the energy, momenta and angular information are measured by the particle detectors. Other fundamental particles such as the tau lepton or the electroweak bosons or the Higgs boson have short lifetimes and decay to the above-listed particles before reaching the detector. As such, these particles can be reconstructed using the information of the measured final state particles. A specific final state or event class refers to a particular combination of final state objects, for instance, a di-muon final state would consist of two muons. There can be additional requirements defined on the object content and on the observables (such as momentum or angle) of such final state objects to define data selection regions where searches for new physics can be concentrated. For any new particle or phenomenon of BSM physics to be seen at such experiments, it is expected that the hypothetical new particles decay into the final state particles that are observed experimentally. Another signature could be that the new physics phenomena affect the cross section of SM physics processes or have a notable effect on the distribution of the kinematic observables of the final state objects that are detected. As a result, anomalies could be seen in the number of events observed in certain data selection regions (or final states) or in features of kinematic observables, in comparison to the SM expectation.

The search strategy is developed based on the assumptions that a signal of new physics can be found as a statistically significant deviation of the event counts in the experimental data from the expectation—in a specific data selection region. The data selection region could refer to any set of requirements on objects or variables needed to define a phase space region where a signal can be searched for, such as a final state or a specific range in one or multiple distributions of observables. A step by step description of the general model independent search approach is described below.

- Event selection: The dataset on which the search is performed is recorded by the experiments during particle collisions using the different trigger algorithms that the experiment might have. While the specific collision events to be used in the search analysis can be selected based on the trigger information and physics object content of the event, general model independent searches tend to be liberal in terms of event selection in order to probe as wide a part of the recorded dataset as feasible. The Standard Model expectation is usually estimated using Monte Carlo (MC) simulations for the different physics processes of the SM. The simulated MC samples are also designed to take into account the detector effects that would be seen in the experimental data in order for the simulated samples to give a more accurate description of what is to be expected in the experimental data. An alternate approach would be to model the background using the experimental data distribution in an alternate data selection region, where it is assumed that there is no effect of BSM physics. Then this distribution is extrapolated into the search region. Since, for a general model independent search approach, it would be preferable to make no assumptions about BSM physics and which selection regions they might influence, the first method of background estimation is the usually preferred method, although in certain implementations of this approach corrections to the simulation are performed based on some input from the data.

- Classification of events: The dataset and simulated MC samples must then be divided and categorised based on the event content, such as photons or electrons, since the different sets of collision events in the dataset are expected to correspond to different final states. Figure 3.1 shows an example classification used by CMS. Here both exclusive and inclusive event classes are used. For a general search, this could mean up to several hundred different final states or data selection regions within which the data and simulated events must be categorised according to their main features. Hence, the event categorisation step would benefit from a high degree of automation, and the experimental implementations of the model independent approach often use software algorithms to automate this step.

- Observables to be probed: After obtaining the different final states with data content and background estimates, comparisons of data and the background are to be made to search for any deviations. The comparisons can be purely in terms of the overall event yields, that is the number of events recorded in the different final states, or using certain kinematical distributions and searching for deviations in the shapes or the yields in a particular part of the phase space in data compared to the background. Certain kinematical distributions are preferred for analyses for which the likelihood of the signatures of a wide range of BSM phenomena being identified are greater, and the most common such distributions are the momentum sum, that is, the sum of the momenta (or transverse momenta) of the physics objects in the final state, the invariant mass of the physics objects in the final state and the Missing Transverse Energy (MET). The motivation for preferring these observables is specified below:

  - The momentum sum distribution is sensitive to high energy BSM phenomena that could show up as a deviation in the tails of the distribution in cases where highly energetic particles with larger momenta are produced compared to the SM expectation.
  - Several BSM theories predict new particles or resonances at higher energies that would decay to the SM particles and such phenomena could be seen as resonant structures in the invariant mass distributions.
  - The MET distribution represents the particles that go undetected in the experiment, and several BSM theories predict such particles, particularly as dark matter candidates. The presence of such particles would lead to a discrepancy in the MET distribution compared to the SM expectation.
  Other distributions, such as the total hadronic activity in the event or the output of a machine learning algorithm, can also be used in such searches.

- Scanning for deviations: The next step is to use a statistical algorithm to scan these data selection regions or each of the kinematic distributions for each final state and quantify the deviations of the data from the SM expectation. The specific algorithm varies based on the specific implementation of this approach in different

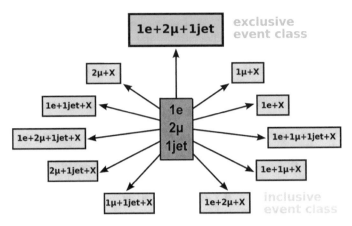

Fig. 3.1 Simplified event classification scheme used by the MUSiC algorithm. Adapted from Ref. [70]

experiments, and some of these are examined in subsequent chapters. In general, such algorithms attempt to identify the data region that has the largest deviation in the distribution of the investigated observable by testing many data selections, and the test statistic, often referred to as the p-value, is calculated for each deviation that can correspond to the expected probability of observing a fluctuation that is at least as far from the SM expectation as the observed number of data. Identification of significant deviations is done by identifying the region with the lowest p-value, that is, the region where the deviation in the data is least likely to be explained by possible fluctuations in the SM background. In many cases, the test statistic used has to be corrected for the trial factor or look-elsewhere effect, which refers to the phenomenon that in a statistical analysis over a large parameter space, seemingly statistically significant observations may arise by chance because of the sheer size of the parameter space explored.

Distributions of kinematic observables, usually in the form of histograms, are investigated for all final states considered in the analysis. Again, due to the sheer number of different final states and selection regions to be considered, this entire process can be automated using a general search algorithm since it is not feasible to develop different methods for each of the different regions explored.

Any statistically significant deviation that might be found is to be studied carefully before claiming that it can be a possible sign of new physics. In particular, it must be verified that the deviation does not fall into the following categories:

- The deviation is not due to an inadequate estimation of the Standard Model prediction. The final state showing the deviation must be checked to ensure that all the possible SM processes that might contribute to such a selection are adequately described and correctly included. This is something that is to be checked in general for the model independent search; however, special attention must be paid to final states showing any deviation.

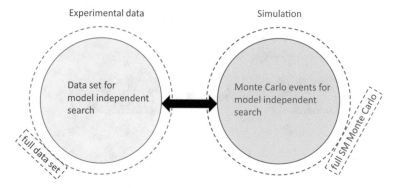

Fig. 3.2 Model independent searches: a large fraction of the full dataset, analysing many different final states, is used to search for deviations from the standard model. The latter is represented by a set of simulated datasets that contains all the final states explored in the search. The comparison uses general algorithms

- The deviation is not due to any misunderstanding or mismodelling of the detector response. Certain final states can be more vulnerable to the effects of the detector response depending on the features of the particle physics detector involved. If any deviation is found in such final states that are vulnerable to be affected by the detector response, it must be carefully examined that the deviation is not due to detector effects.
- The deviation is not a statistical fluctuation. This could be the case if there is a deviation in a final state with low statistics, with only one or a few events corresponding to the deviation. Such cases must be examined carefully to consider the possibility of the deviation being a statistical fluctuation rather than a sign of new physics.

After running through all the steps outlined above, if data selection regions are identified that show statistically significant deviations in the data compared to the SM expectations, then such deviations can be explored and interpreted in terms of possible new physics phenomena.

Figure 3.2 illustrates the use of experimental and simulated datasets in model independent searches.

3.2 Comparison with Dedicated Searches

Dedicated searches are those search analyses that target the signatures of one or a few specific BSM models. They concentrate the search in one or a few limited number of final states where the signature of the BSM model(s) being probed are predicted to appear in. Such searches often use specific selection criteria that are

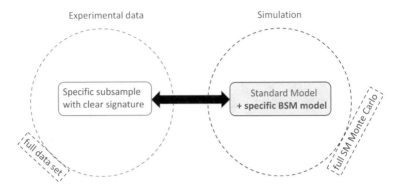

Agreement with prediction by Standard Model + specific BSM model ?

Fig. 3.3 Dedicated searches: only a small part of the full dataset with one or few final states is used to search for new physics predicted in a specific BSM model. The experimental data are compared to the prediction of SM plus BSM physics. The comparison is optimised for the signal under study

designed to suit the features of the specific BSM signatures being probed, and these sometimes use more sophisticated ways to estimate the background in the limited number of final states probed, such as using data-driven methods. These searches are not bound by the limitations of the general model independent searches, where the same approach is applied to numerous different final states and the search algorithm is designed to probe general features that would cover a wide range of possible BSM phenomena, not specifically designed to exploit particular BSM signatures. Certain dedicated searches also provide a model independent interpretation, by comparing the observables in the data to the SM expectation in the selection region or final state probed; however, they still probe only one or a few final states or selection regions and are thus significantly different from general model independent searches.

Figure 3.3 illustrates the use of experimental and simulated datasets in model independent searches.

Several such dedicated searches have been and continue to be performed at the different collider-based high energy physics experiments. However, due to practical constraints in terms of person-power and resources available at experiments, a limited number of such searches can be performed, thus only probing a limited set of BSM models and final states associated with them, often concentrating on the more popular BSM models at the time.

Given the contrast in the two approaches, of the dedicated searches and the general model independent searches, a comparison of the advantages and disadvantages of the two approaches is described below.

3.2.1 Advantages of General Model Independent Searches

The main advantages that general model independent searches have are listed below.

- Dedicated searches are conducted for a limited number of popular BSM models due to practical constraints on the number of searches that can be conducted, thus leaving several other BSM models unexplored particularly since different BSM models often have signatures that are very different from one another. A general model independent search is designed to have a broad scope, covering several different final states or data selection regions. Signs of new physics might be hidden in regions of the experimental dataset that are not investigated by dedicated searches, and using a general search reduces the likelihood of missing any new physics phenomena that might be hidden in the more obscure parts of the recorded experimental dataset.
- Furthermore, there is the possibility that none of the currently hypothesised BSM models describe nature accurately and new physics beyond the SM could have a signature that does not correspond to any of the currently hypothesised BSM models. Such signs of new physics are unlikely to be uncovered by dedicated searches, whereas general model independent searches can find unexpected signals of new physics due to the large number of event classes and phase space regions probed, which may otherwise remain uninvestigated. It is not practically feasible to have dedicated searches for each of the different data selection regions that can be probed using the general approach.
- Even in final states probed by dedicated searches, often the search analyses are designed to look for specific features corresponding to the particular BSM model being probed. Thus, signs of new physics phenomena showing different characteristics compared to that being probed by the dedicated search could be missed. A model independent approach that looks only for deviations from SM expectation instead of particular characteristic signatures of a specific BSM model has the ability to uncover any deviating features in the dataset where the features of such deviation are different from that probed by the dedicated search.
- Another advantage of the general model independent search approach is that it could identify several small deviations in multiple similar final states, that could hint at something new when considered in combination. Such effects would be missed by dedicated searches, where relatively small deviations might be overlooked without the benefit of a larger picture.
- Aside from the main objective of searching for new physics phenomena, the general model independent search approach provides a comparison between the recorded dataset and the background expectation that is generally obtained from the simulation of SM processes. Since the approach is broad, the scanned distributions can be used to probe the overall description of the data by the event generators that are used to simulate the many SM processes and also validate the overall experimental understanding of the detector related effects in a large number of different selection regions.

- The implementations of the general model independent search approach at certain experiments also include the usage of pseudo-experiments to evaluate the probability of a deviation occurring in any of the many different final states under study. This is done in order to understand the deviations in the data that might occur purely due to chance given the large number of final states probed. The benefit of such an evaluation is that it would result in a global interpretation of the probability of finding a deviation within an experiment, possibly providing a context even for the several different independent dedicated searches performed at the experiment.

3.2.2 Disadvantages of General Model Independent Searches

Along with the advantages, there are also some drawbacks for general model independent searches. The major disadvantages are listed below.

- A general model independent search is not optimised for a specific class of BSM signals, and a dedicated analysis optimised for a given BSM signal can achieve a greater sensitivity to that signal. The optimisation can be in terms of the particular event selection, or in terms of the specific or combination of observables probed. The enormous parameter space of possible signals makes an optimised search for each of them impossible, and special variables or selection criteria that might increase the sensitivity for particular BSM models do not make sense for a general search.
- Dedicated analyses probing a particular model can, for the case where no signs of signal are found over the estimated background, place limits on the masses or cross sections for the hypothesised particles for the BSM model. This helps to rule out parameter space that has already been explored in terms of particular BSM models for future searches or theoretical development. General searches only look at deviations from the SM and do not endeavour to set limits or rule out parameter spaces for BSM models.
- In general, model independent searches depend on Monte Carlo simulations for the description of physics processes and simulations of the experimental effects and detector response. Event classes in which the majority of the events contain mis-reconstructed objects are typically poorly modelled by MC simulation and might need to be excluded from the analysis. This leaves the analysis potentially vulnerable to MC mismodelling in a certain corner of the phase space regions probed. This could be improved if a better background model could be constructed with the help of control regions or data-derived fitting functions, but this might make the analysis much more complex due to the diverse different final states that need to be described. Since dedicated searches only probe restricted data selection regions, sometimes such searches use the data distributions in other so-called 'control regions' to derive a data-driven background estimation for their signal region where the search is performed.

- Although general model independent searches are very broad in terms of the data selection regions and phase space probed, the procedure might still miss a certain BSM signal, in case it does not show a localised excess in one of the studied kinematic distributions. One way to overcome this would be to probe a wider range of observables and use modified observables; however, it might make the analyses much more complicated and less practical. In addition, the statistical trial factor will increase.
- Since general model independent searches have a wide scope, probing numerous different final states, they also require a complex software framework to execute the idea and more computing resources to run it as compared to simpler dedicated analyses. As such, adding further complexities to the analysis strategy can be difficult. For such reasons, the timescale of a general model independent search analysis can be longer than that of simpler dedicated analyses.
- The sensitivity of a general model independent search may be reduced due to the statistical trial factor, which has to be considered. In most cases, model independent searches have a large number of search regions, and its reduction would imply a loss of generality. While dedicated searches for new physics also need to account for a trial factor, it can be small or negligible, e.g. when searching for the manifestation of a model with a very specific prediction.

3.2.3 Complementary Nature of General Model Independent Searches and Dedicated Searches

While the previous sections provide a comparison between general model independent searches and dedicated searches for particular BSM models, presenting the advantages as well as the disadvantages, it is not necessary to look at the two approaches as competing exclusive approaches, but rather it can be more beneficial to consider the two approaches as complementary to one another in the context of experimental searches for BSM physics.

BSM physics models with unusual signatures or signatures in specific kinematic distributions that are not suitable for general model independent searches must be probed with dedicated searches. This is also valid for searches targeting signatures in final states where there is likely to be background from sources that are not easily modelled by simulation, such as beam halo, where the model independent approach might not be adequately sensitive. A dedicated data-driven background estimation might be required for such cases that are more suitable for dedicated searches. The experiments have several dedicated searches targeting the more popular BSM models of the day; however, this has limited scope in terms of the phase space regions and BSM models probed.

The case for the general model independent search has already been made, to target phase space areas and BSM models uncovered by the dedicated searches since there are practical limitations on the number of feasible dedicated searches. Even in final states covered by dedicated searches, in case the dedicated search probes a specific signature and thus becomes likely to miss a different signature, the general search becomes useful. Conversely, if a general search does reveal a discrepancy in a particular final state, it might be useful to have a dedicated search probing that final state in detail, to understand the discrepancy better, and to compare it with possible signatures of likely BSM models that might be able to explain such a discrepancy. In case general searches see smaller deviations, a dedicated search analysis could show a clearer picture.

The ultimate goal of both approaches is to conduct experimental searches of new physics beyond the SM and having a complementary strategy of dedicated searches and general model independent searches would be more suitable to that end. Taken together, these contrasting approaches would enable the probe of wide regions of the phase space, while at the same time not compromising on sensitivity of searches. All of the high energy physics experiments where a general model independent search has been performed also have a programme consisting of several dedicated search analyses, thus effectively benefiting from the advantages of both approaches.

Specific implementations of the general model independent search approach vary according to the different experiments where such searches have been conducted. It is very instructive to discuss the specific instances of the model independent searches that have been conducted. The following chapters describe general model independent searches as implemented at different high energy physics collider experiments.

Chapter 4
General Model Independent Searches in Past Collider-Based Experiments

General model independent search analyses have been performed at collider experiments that were running in the past. This chapter focuses on the implementation of the general model independent search strategy in such analyses, particularly on the analyses conducted at the CDF and D0 experiments at the Tevatron and also at the H1 experiment at HERA.

4.1 Model Independent Searches at the Tevatron

The Tevatron collider at Fermi National Accelerator Laboratory, USA was a proton–antiproton collider running at a center-of-mass energy of up to 2 TeV that was in operation between 1983 and 2011. The major collider experiments at the Tevatron were the CDF [59] and the D0 [60, 61] experiments. These experiments are well known for the first experimental discovery of the top quark [62, 63], and they also had an extensive physics programme aimed towards searching for BSM physics, including general model independent search analyses. The general model independent searches conducted by the D0 and the CDF experiments are described below, further details of these analyses can be found in Refs. [71–74] for the searches conducted at the D0 experiment, and in Refs. [75, 76] for the searches conducted at the CDF experiment.

The strategy employed in the most recent implementation of the general model independent search approach by the experiments based at Tevatron is discussed here. Two complementary search algorithms have been implemented, one that performs a systematic and model independent search looking at gross features of the data (VISTA), and the other being a quasi-model-independent search for new physics at high transverse momentum (SLEUTH). Furthermore, CDF also employed the BumpHunter algorithm [76], which is not discussed here. In the following, first general features pertaining to both the algorithms and their implementation at the

© The Author(s), under exclusive license to Springer Nature Switzerland AG 2020
S. S. Ghosh et al., *General Model Independent Searches for Physics Beyond the Standard Model*, SpringerBriefs in Physics,
https://doi.org/10.1007/978-3-030-53783-8_4

two experiments are discussed in Sect. 4.1.1, followed by descriptions of the VISTA and SLEUTH algorithms in Sects. 4.1.2 and 4.1.3, respectively. Finally the specifics pertaining to the implementation at the D0 and CDF experiments are described in Sects. 4.1.4 and 4.1.5, respectively.

4.1.1 General Features

The VISTA algorithm aims to perform a global study of the Standard Model prediction compared to the observed data in the bulk of the high transverse momentum data in order to search for any discrepancies that could be explored as signatures of new physics phenomena, and the SLEUTH algorithm complements this with a search for possibly small cross-sectional physics in the tails of the regions with high transverse momentum. The two approaches are employed together for the general model independent search for new physics by the CDF and D0 experiments at the Tevatron. Hence, they share some general features that are described here before discussing each of the two approaches separately.

The Standard Model (SM) processes that form the background are estimated using simulated samples obtained by generating a large sample of Monte Carlo events. The detector response of the CDF or the D0 detector, as the case may be, are taken into account during the simulation process. The VISTA algorithm also includes a derivation of certain correction factors for the simulation.

For both VISTA and SLEUTH approaches, events containing an energetic and isolated electron, muon, tau, photon or jet objects were selected. Based on these objects that are considered, experimental data and simulated Standard Model events are categorised into exclusive final states. This categorisation is done in an orthogonal manner, with each event ending up in one and only one final state. Data are compared to the Standard Model prediction in each final state, to search for statistically significant deviations using the VISTA and SLEUTH algorithms.

The final states subjected to the SLEUTH algorithm by the D0 experiment [72] are summarised in Fig. 4.1.

This search for new physics terminates either when one or more significant deviations are found to present a compelling case for new physics, or when there remain no statistically significant discrepancies which can be probed for new physics phenomena. For the first scenario, it would be essential to quantitatively assess the significance of the potential discovery, whereas for the latter case, it is sufficient to demonstrate that all observed effects are not in significant disagreement with an appropriate global SM description based on an appropriate quantitative threshold on what constitutes a statistically significant deviation.

Fig. 4.1 Event classes analysed by the D0 collaboration using the SLEUTH algorithm [72]. Each row in the diagram represents one object class, empty rows denote jets. For example, the left lower part of the figure refers to event classes with one electron, missing transverse energy plus one, two, three, four or more jets. Rows marked by a blue filled circle have been analysed with SLEUTH, the green triangles indicate earlier D0 searches

4.1.2 VISTA

The VISTA approach probes deviations that would appear in the bulk of the distributions rather than narrow regions of phase space. This includes examining the overall agreement between data and simulation in terms of total yield or normalisation along with evaluating the overall compatibility of complete kinematic distributions between the data and simulation. However, such probes are not the only features of the VISTA approach, and deriving correction factors in order for the simulation to better describe the data forms an important aspect of the analysis. The VISTA approach is described below in two parts, first describing the method that is used to obtain correction factors, and then discussing the examination of discrepancies between data and simulation in the bulk of kinematic distributions in different final states.

VISTA correction factors: The VISTA approach deals with the shortcomings that the simulation might have in terms of describing the data that are not linked to new physics phenomena and that cannot be determined from first principles, through correction factors obtained using the Vista correction model. This correction model is applied to the simulation to obtain the Standard Model prediction across all final states. Correction factors are to be obtained from the data and most analyses that are dedicated to searching in specific final states use control regions in data where no signal is expected to derive corrections. This procedure is not compatible with a general model independent search strategy. The approach adopted instead is to obtain a consistent set of correction factors across the wide range of final states considered.

The correction factors are classified into two categories: theoretical and experimental. Theoretical correction factors take the form of k-factors, which correspond to the ratio of the unavailable higher order cross-sectional calculations to the cross sections calculated at leading order. The k-factors are used for Standard Model processes including QCD multijet production, W+jets, Z+jets, and (di)photon+jets production. Experimental correction factors include the integrated luminosity of the data, efficiencies associated with triggering on and the correct identification of physics objects, and misidentification rates associated with the mistaken identification of physics objects.

The procedure to obtain the correction factors involves an iterative process of improving the simulation based on the inadequacies in the modelling of data by simulation. A major concern of such a process of obtaining correction factors would be that the correction factors obtained based on observed discrepancies may allow a real signal of BSM physics to be artificially suppressed. While it would be the case for any analysis, that if adjusting the correction factors within allowed bounds removes a deviation, then the deviation can be said to be explained in terms of known physics or detector phenomena. It must be ensured, however, that there are strong constraints placed on the correction model so as to avoid missing any real signs of new physics phenomena. The VISTA correction model accounts for this by requiring the correction factors to provide a consistent interpretation of hundreds of different final states simultaneously, thus making it much less likely to mistakenly explain away new physics than if the approach was confined to a more limited scope.

The values of the correction factors are obtained from a global fit to the data. Events are categorised into final states based on the number and types of objects and then further divided into bins based on the transverse momentum and pseudorapidity of the objects. The simulated events that provide the Standard Model expectation are adjusted based on the correction factors, thus the total simulated expectation in each bin can be taken as a function of the values of the correction factors. The correction factors are then optimised to give the best agreement between data and simulation across all the different bins by using the figure of merit that is given in Eq. (4.1). In the equation, $\vec{s}$ represents a vector of correction factors for the kth bin, $Data[k]$ is the number of data events observed in the kth bin, $SM[k]$ is the number of events predicted by the Standard Model in the kth bin, $\delta SM[k]$ is the statistical uncertainty on the simulated SM prediction in the kth bin and $\sqrt{SM[k]}$ is the statistical uncertainty on the expected data yield in the kth bin. The SM prediction, $SM[k]$, is a function of the correction factors, $\vec{s}$, and the values of the correction factors are obtained from the global fit across bins and final states providing the best global agreement between the data and the SM prediction. It must be noted that some of the correction factors are also constrained by additional information regarding higher order theoretical calculations or measured detector effects.

$$\chi_k^2(\vec{s}) = \frac{(Data[k] - SM[k])^2}{\sqrt{SM[k]}^2 + \delta SM[k]^2} \tag{4.1}$$

What is discussed here is the basic outline of the VISTA correction procedure, and further details can be found in Ref. [76].

Once the correction factors have been evaluated, the simulated SM expectation is then obtained using all the correction factors. This corrected SM expectation is then used in the further steps of the analysis to compare with the observed data in order to probe for any deviations that could be interpreted as signs of new physics phenomena.

VISTA search approach: The VISTA search approach examines discrepancies that affect the overall distributions, particularly dealing with the overall numbers of expected events and simulation to data agreement across complete distributions of kinematic variables. This search is performed in the exclusive final states into which the data and simulated SM events, after including the correction factors, are categorised into based on the object content. The categorisation is performed in an orthogonal manner such that each event belongs to only one final state.

The VISTA method to probe for discrepancies between the simulated SM expectation and the data is made up of two parts. First, the normalisation is checked based on the total number of events in each exclusive final state by using a measure of the goodness of fit that is calculated using Poisson probabilities. Then, the overall consistency between the data and estimated SM distribution for a number of kinematic distributions is checked using a Kolmogorov–Smirnov (KS) statistic and the resulting fit probability.

Before getting into the details of the evaluation of the deviations, it is important to address the issue of the look-elsewhere effect or the trial factor quantifying the number of places an interesting signal could appear, considering that, when such a large number of regions are scanned in an algorithm like VISTA, a certain number of fluctuations with statistically significant deviations are expected to appear simply due to the large number of regions being considered. When observing many final states, some disagreement is expected from statistical fluctuations in the data. This effect is accounted for by modifying the statistic that is used to measure the deviations (the Poisson probability for the case of comparing total event yield and the Kolmogorov–Smirnov statistic for the case of the comparison of distributions) based on the knowledge of the full scope of the regions that are analysed within the VISTA approach. Finally, a discrepancy at the level of 3σ or greater after accounting for the trial factor is considered 'significant', with that value typically corresponding to a discrepancy at the level of 5σ or greater before accounting for the trial factor.

As indicated before, the discrepancy in the total number of events in each final state is measured by the Poisson probability taking into account the number of predicted events that would fluctuate upwards or downwards to cover the number of events observed. In order to account for the trial factor or look-elsewhere effect associated with the number of final states probed, the VISTA algorithm uses another statistic for each final state probed, that is the $p\,value$, shown in Eq. (4.2), where p_{fs} is the Poisson probability in a particular final state, and N_{fs} is the number of final states probed.

$$p = 1 - (1 - p_{fs})^{N_{fs}} \qquad (4.2)$$

The obtained value of p can then be converted in terms of the units of standard deviation by solving for the term σ according to Eq. (4.3), and a final state with a deviation of larger than 3σ can be considered significant.

$$\int_{\sigma}^{\infty} \frac{1}{\sqrt{2\pi}} e^{-\frac{x^2}{2}} dx = p \tag{4.3}$$

Apart from the comparison of total event yields in each final state, the VISTA approach also compares the simulation expectation to the observed data distribution in the bulk of several kinematic distributions in each final state. Several kinematic distributions are considered for each final state, including the transverse momentum, pseudorapidity, and azimuthal angle of all objects, masses of individual jets and b-jets, invariant masses of all object combinations, transverse masses of object combinations including the missing momentum, angular separation $\Delta\phi$ and ΔR of all object pairs, and several other more specialised variables. To evaluate quantitatively the overall difference in the shape of each kinematic distribution between data and SM prediction, a KS test is used. Again, as for the case of the comparison of total yields, the trial factor is evaluated to take into account the numerous kinematic distributions that are scanned, and the resulting probability value is then converted into units of standard deviations. A distribution with KS statistic greater than 0.02 and probability corresponding to greater than 3σ after assessing the trial factor is considered significant.

This general analysis to search for discrepancies between the observed data and SM expectation obtained from simulation in the total yield and overall shapes of kinematic distributions in several different finals based on the VISTA approach is complemented by a search for deviations in regions corresponding to high transverse momentum using the SLEUTH algorithm.

4.1.3   SLEUTH

The SLEUTH algorithm is complementary to the VISTA algorithm and is specifically designed to search for deviations between the data and the simulated SM prediction in events at high transverse momentum (p_T). SLEUTH can be referred to as a 'quasi' model independent search, where the term 'quasi' is used since SLEUTH relies on the assumption that the first sign of new physics will appear as an excess of events in some final state at large summed scalar transverse momentum (p_T). This is a well-motivated assumption since when considering a broad range of BSM physics models, an interesting commonality can be noted that is shared by a large number of BSM models, which is that they predict an excess of events at high p_T, concentrated in a particular final state.

The basic assumptions behind the SLEUTH approach are: first, the data can be classified into exclusive final states such that the signs of new physics will show up predominantly in one of the final states. Second, the signature of new physics will appear in the form of objects at high summed transverse momentum ($\sum p_T$), and in such kinematic regime, the data will show deviations compared to the SM expectation. Third, the signature of new physics will appear as an excess of data over the SM and instrumental background.

The SLEUTH algorithm is described in brief below, describing the three basic steps that make up the algorithm. Further details can be found in Refs. [71–74, 76].

1. First, the events are categorised into exclusive final states, as for the VISTA algorithm, but with a few modifications to make the analysis simpler, as listed below.

   - In final states with jets, in the CDF analysis jets are grouped together in pairs [76]. This reduces the total number of final states and also keeps signal events with one additional radiated gluon, that is presumably the source of the jet, within the same final state. In cases with an unpaired jet, it is assumed to have originated from a radiated gluon, and not an individual final state object.
   - The same method as for the case of jets is applied to b quark jets.
   - Final states that can be considered as global charge conjugates are taken together. For instance, final states with two electrons (e) and a photon (γ) can be: $e^+e^+\gamma$, $e^+e^-\gamma$ and $e^-e^-\gamma$, and for these the states $e^+e^+\gamma$ and $e^-e^-\gamma$ are clubbed together as a single SLEUTH final state.
   - Final states with interchange of the first and second generation leptons are considered to be equivalent. For instance, the final state with one muon (μ) and a photon, $\mu^+\gamma$, is clubbed together with the $e^+\gamma$ final state.

2. Second, the variable considered for probing for deviations in this algorithm is a single variable in each exclusive final state: the summed scalar transverse momentum of all objects in the event $\sum p_T$. This also includes, along with the momenta of all the reconstructed objects, the missing transverse momentum and also the vector sum of all transverse momentum visible in the detector but not clustered into an identified object.

3. Third, the algorithm then searches for regions in which more events are seen in the data than expected from the Standard Model and instrumental background in the variable space defined in the second step of the algorithm, for each of the exclusive final states defined in the first step. The regions are evaluated using the following steps:

   - In each final state, the regions considered are one-dimensional semi-infinite intervals in $\sum p_T$ extending from each data point to infinity, with the requirement that the region must contain at least three data events.
   - A Poisson probability is calculated for each region in the following way. If the data points in a final state are ordered in reverse by $\sum p_T$, then the region beginning with the dth data point contains d data events and the SM expectation is estimated for the region based on simulated events corrected with the VISTA method, by integrating the simulation estimates yield in that region to the value, let us say, of b predicted events. Then, in this final state, the Poisson probability for the dth region is given by p_d as described in Eq. (4.4). The region with the smallest p_d value is considered to be the most interesting region in that final state.

$$p_d = \sum_{i=d}^{\infty} \frac{b^i}{i!} e^{-b} \qquad (4.4)$$

- The correction for the trial factor or look-elsewhere effect is performed using pseudo-experiments that are generated with pseudo-data pulled from the SM background. For each pseudo-experiment, the most interesting region is calculated. Using an ensemble of such pseudo-experiments, the fraction $\mathcal{P}$ of pseudo-experiments with a smaller Poisson probability for the most interesting region than that observed in data for a particular final state is evaluated. The value of $\mathcal{P}$ is expected to be small for the case where there is a significant deviation observed in data possibly coming from a signal of new physics. Otherwise, it is expected to be a random number pulled from a uniform distribution in the unit interval. This process is done in each of the final states. The minimum of these values is denoted $\mathcal{P}_{min}$ and the most interesting region in the final state with smallest $\mathcal{P}$ is denoted $\mathcal{R}$.
- A statistic referred to as the interest of the most interesting region $\mathcal{R}$ in the most interesting final state is defined by $\tilde{\mathcal{P}}$ as in Eq. (4.5), where the product is over all SLEUTH final states a, and $\hat{p}_a$ is the minimum of $\tilde{\mathcal{P}}_{min}$ and the probability for the total number of events predicted by the SM in the final state a to fluctuate up to or above three data events.

$$\tilde{\mathcal{P}} = 1 - \prod_a (1 - \hat{p}_a) \qquad (4.5)$$

$\tilde{\mathcal{P}}$ represents the fraction of hypothetical cases that would produce a final state with $\mathcal{P} < \mathcal{P}_{min}$. In case the data distribution can be explained by the SM expectation, the $\tilde{\mathcal{P}}$ value would be expected to be a random number pulled from a uniform distribution in the unit interval. On the other hand, if signature of new physics is present, then $\tilde{\mathcal{P}}$ is expected to be small.

4. Finally, the output of the SLEUTH algorithm is the most interesting region observed in the data ($\mathcal{R}$), and the value quantifying the interest of $\mathcal{R}$, that is $\tilde{\mathcal{P}}$. For the SLEUTH algorithm, a reasonable threshold for discovery is taken to be $\tilde{\mathcal{P}} \lesssim 0.001$, which corresponds to a deviation with local significance of a 5σ effect after the trial factor is accounted for.

Having described the basic outline of the SLEUTH search algorithm, it must also be mentioned that the SLEUTH algorithm has been tested for its discovery potential. Such tests have been performed using two approaches, first by removing a known Standard Model process from the background estimation and testing SLEUTH's ability to discover the missing process, and second by first generating pseudo-data adding the expectation of a certain benchmark BSM scenario to the simulated SM expectation and then again testing the ability of SLEUTH to detect this.

   One of the tests performed by removing a known Standard Model process from the background expectation was done by removing the top quark pair production

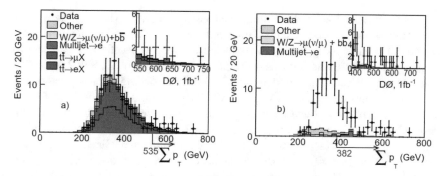

Fig. 4.2 D0 analysis of final states with a large contribution from $t\bar{t}$ events [74]. The left figure shows the comparison to the SM with top contribution, in the right figure the top has been removed from the simulated data sample. The inserts show the high momentum tails of the sum of the transverse momenta distributions

process from the SM expectation to quantitatively test SLEUTH's sensitivity to top quark pair production. After removing this process from the SM expectation, the values of the VISTA correction factors were re-obtained from a global fit. Then the SLEUTH algorithm was used based on the conditions for a dataset with an integrated luminosity corresponding to the analysis using the final CDF dataset. It was found that the SLEUTH algorithm would be able to detect the missing top quark pair production process with a significance far exceeding the discovery threshold mentioned before. Similar checks were performed for the analysis at the D0 experiment. Figure 4.2 demonstrates how the top quark could be (re-)discovered using the SLEUTH algorithm [74]. The D0 collaboration examined the final state with a light charged lepton, four jets, two of them b-jets, plus missing transverse energy and analysed the distribution of the sum of the transverse momenta of all objects on one event. Clearly, top pair production is needed to describe the measurements.

Multiple tests were performed using benchmark BSM models such as a Gauge Mediated Supersymmetry Breaking (GMSB) scenario or the signal with a hypothetical W' boson, and for several of the cases, the SLEUTH algorithm was found to perform comparably with respect to the dedicated analyses targeting such BSM models.

4.1.4 Implementation at the D0 Experiment

The most recent analysis by the D0 experiment, described in Ref. [74], was performed on a 1.1 fb^{-1} dataset collected by the D0 detector in proton–antiproton collisions at a center-of-mass energy of 1.96 TeV at the Tevatron collider. While the general strategies of VISTA and SLEUTH algorithms were adopted, some modifications were made.

The analysis focused on final states that contained leptons, with only those events with at least one electron or muon being considered. Specific identification criteria were used to select energetic objects isolated from other event activity, viz., electrons, muons, tau leptons, missing transverse energy, jets and b-quark jets. Events with photons were rejected, mainly due to difficulties in modelling using simulation.

Before analysing the events with the VISTA and SLEUTH algorithms, first the data and the selected Monte Carlo simulated events were divided into seven inclusive subsets based on the number and types of leptons identified in each event and for each of these, corrections were applied to the simulation obtained from previous D0 studies based on well-understood regions of phase space, dominated by particular SM processes. The inclusive non-overlapping subsets correspond to the single electron and jets, single muon and jets, double electron, double muon, electron and muon, electron and tau lepton and muon and tau lepton inclusive final states with additional criteria to ensure that the states are non-overlapping. To account for any incorrect normalisations, fits were performed for contributions from each of the subsets to obtain scale factors which reproduce the distributions in the selected data with simulated events and multijet background events determined from data.

Then, the seven non-overlapping inclusive subsets were merged to create an input file for the analyses employing the VISTA and SLEUTH algorithms that have been described before, with the analysis then being performed on exclusive final states based on the object contents of the events as described before while describing the VISTA and SLEUTH algorithms.

The analysis with the VISTA algorithm using the D0 dataset resulted in 117 different exclusive final states being probed. The comparison of the total event yields in each of the final states showed that only two out of the 117 exclusive final states had statistically significant deviations, and both of those deviations were studied and attributed to problems with modelling the data with simulation rather than signs of new physics phenomena. The two discrepant final states were the single muon, two jets and missing transverse energy final state and the two muon and missing transverse energy final state. The discrepancy in the first final state was attributed to likely problems with modelling jets associated with initial state radiation and final state radiation in the forward regions of the detector, while the second discrepancy was attributed to difficulties modelling the muon momentum distribution for muons with high transverse momenta. The analysis with the VISTA algorithm probing the overall shape for different kinematic distributions involved probing a total of 5543 individual one-dimensional distributions in various variables in the 117 exclusive final states. Of these, sixteen distributions were found to be discrepant but each of the discrepant distributions was found to be related to known simplifications in the modelling assumptions.

As an example Fig. 4.3 shows two kinematical distributions in the muon plus jet plus missing transverse energy event class, analysed with VISTA by the D0 collaboration [74].

The analysis with the SLEUTH algorithm was performed in 31 final states, into which the 117 final states of the VISTA analysis were combined based on the criteria described in Sect. 4.1.3. The SLEUTH analysis also did not find any final states that

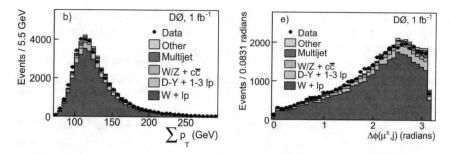

Fig. 4.3 VISTA analysis of events with muon plus jet plus missing transverse energy, taken from Fig. 13 in Ref. [74]. The left figure shows the distribution of the scalar sum of the transverse momenta, in the right part the distribution of the angle between muon and jet directions is displayed. Overall there is a good agreement between D0 data and SM prediction. The small discrepancies are attributed to systematic uncertainties in the modelling, which are not taken into account when calculating significances

Fig. 4.4 SLEUTH analysis of events with light charged lepton (e, μ) plus tau plus missing transverse energy [74]. The insert shows the high tail of the distribution

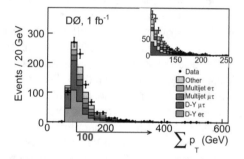

exhibited significant deviations which could not be explained by the shortcomings in the simulation.

Figure 4.4 shows the distribution of the scalar sum of transverse momenta in events with an electron or muon, plus a tau and missing transverse energy [74], exhibiting a statistical discrepancy with a significance of about 3 σ.

The analysis with the D0 dataset concluded that all observed discrepancies that were found between data and simulation could be attributed to uncertainties in the Standard Model background modelling, and hence no evidence for physics beyond the Standard Model was found based on this analysis.

4.1.5 Implementation at the CDF Experiment

The general model independent search performed most recently by the CDF experiment based on the VISTA and SLEUTH algorithms was performed on a dataset collected by the CDF detector corresponding to 927 pb^{-1} in proton–antiproton collisions at a center-of-mass energy of 1.96 TeV at the Tevatron collider [76]. The analysis searched over three hundred different exclusive final states for deviations between data and the simulated SM expectation.

The analysis was done largely following the VISTA and SLEUTH algorithms described in Sects. 4.1.2 and 4.1.3, respectively. Events were selected containing energetic and isolated electrons, muons, taus, photons, jets and b-tagged jets. The VISTA correction model was used to obtain 44 different correction factors. Corrections related to object identification efficiencies were typically less than 10%; correction factors for the misidentification rates were found to be consistent with an understanding of the underlying physical mechanisms responsible, and the k-factors for the normalisation of different physics processes range from slightly less than unity to greater than two for some processes with multiple jets. The values obtained were found to be physically reasonable.

The analysis with the VISTA algorithm led to the examination of 344 different final states, most of which had not been part of previous analyses at CDF. The probe for differences between the total number of observed and predicted events did not yield any statistically significant deviations after accounting for the trial factor. For the step quantifying the difference in shape of kinematic distributions using the Kolmogorov–Smirnov statistic, a total of 16,486 distributions were considered. A few of the distributions did show significant discrepancies between data and Standard Model prediction; however, these discrepancies were believed to arise from mismodelling of the SM prediction, particularly of the parton shower process and the overall transverse boost of the physics interaction during the collision. Thus, none of the discrepancies observed could motivate a claim for new physics phenomena.

Figure 4.5, taken from Ref. [76], shows a prominent discrepancy between CDF data and the SM prediction. The figure shows for three-jet-events the distribution of the angular separation of the two jets with the smallest p_T values. In Ref. [75] this discrepancy is discussed in detail and many indications for a mismodelling of the QCD simulation programmes are put forward. Therefore a new physics claim cannot be derived.

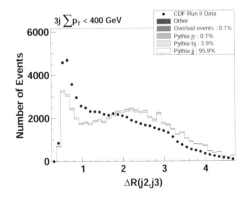

Fig. 4.5 VISTA analysis by the CDF collaboration of events with three jets [76]. Shown is the distribution of the angular separation of the two jets with the lowest transverse momenta

Fig. 4.6 SLEUTH analysis
by the CDF collaboration of
events with two charged
leptons (electron or muon) of
same sign [76]

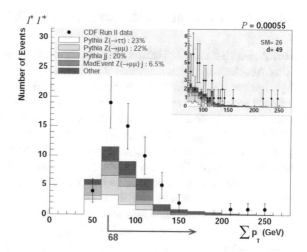

Fig. 4.6 SLEUTH analysis by the CDF collaboration of events with two charged leptons (electron or muon) of same sign [76]

The analysis with the SLEUTH algorithm was also conducted to search for regions of excess on the high-p_T tails of exclusive final states. Many of the exclusive final states searched with the SLEUTH algorithm had not been considered by previous CDF analyses. Following the SLEUTH approach, the measure of interest was calculated for many regions in the final states considered after accounting for the trial factor associated with looking at many such regions in order to quantify the most interesting region observed in the CDF data. After the statistical analysis, no region of excess on the high-p_T tail of any of the SLEUTH exclusive final states was found to surpass the discovery threshold.

Figure 4.6 shows the findings of the CDF SLEUTH search for the event class with two charged light leptons of the same charge [76]. While the P-value is low (0.00055), the significance is nevertheless small, when taking the look-elsewhere effect into account, yielding $\tilde{P} = 0.08$.

The general model independent search performed by the CDF experiment probed several hundred different final states, many of which were not part of previous analyses, to search for statistically significant deviations between the data and simulated Standard Model expectation. The analysis of the CDF dataset did not find any statistically significant deviations that could not be explained by the inadequacies in the modelling of the Standard Model expectation. Thus, no signatures of new physics phenomena were found by this analysis.

Although both the CDF and the D0 general model independent search analyses did not find signatures of new physics phenomena, these broad searches of the datasets collected at the Tevatron represented two of the most encompassing tests of the Standard Model of particle physics at the energy frontier at the time of their publication.

4.2   Model Independent Searches at HERA

HERA, which stands for Hadron Elektron Ring Anlage in German (Hadron-Electron Ring Accelerator in English), was a particle accelerator at Deutsches Elektronen-Synchrotron (DESY) in Hamburg in Germany. At HERA, electrons or positrons were collided with protons at a center-of-mass energy of up to 319 GeV. HERA began its operations in 1992 and was closed down in 2007. During its period of operation, HERA was the only high energy lepton–proton collider in the world. The lepton–proton interactions provide a testing ground for the Standard Model complementary to lepton–lepton colliders and hadron colliders (such as the Tevatron or the LHC).

There were four large particle physics experiments located at the four interaction regions where the particle collisions took place. These experiments were the H1, ZEUS, HERMES and HERA-B experiments. The H1 and ZEUS experiments had multipurpose particle detectors and had the physics goals of investigation of the internal structure of the proton through measurements of deep inelastic scattering, studies of the fundamental interactions between particles in order to test the Standard Model of particle physics, as well as searching for new physics beyond the Standard Model. Both of these experiments went into operation in 1992. The HERMES experiment started taking data in 1995 using the HERA electron beam to investigate the intrinsic angular momentum of protons and neutrons. The HERA-B experiment ran from 1999 to 2003, and its main objective was to study the properties of heavy quarks.

The experiments at HERA produced important studies on the structure of the proton. Apart from this, the H1 and the ZEUS experiments also had a physics programme which included searches for new physics phenomena. Analyses with the approach of a general model independent search for new physics were conducted at the H1 experiment. The general model independent search performed at the H1 experiment is described here in further detail.

4.2.1   Model Independent Search at the H1 Experiment

The H1 experiment was one of the major particle physics experiments with a multipurpose particle detector at HERA in DESY, Hamburg. The H1 detector began operating together with HERA in 1992 and took data until 2007. The physics programme of the H1 experiment constituted studies of using deep inelastic scattering measurements to study the internal structure of the proton, performing measurements of further cross sections that can shed light on fundamental interactions and act as a test of the Standard Model, and to perform searches of new physics beyond the Standard Model. Further details of the H1 experiment can be found in Ref. [64].

General model independent searches for new physics were conducted at the H1 experiment, and the results wer epublished in 2004 [77] and in 2009 [78], using a larger dataset. The search algorithm used is described below and is similar in some ways to the SLEUTH algorithm employed by the D0 and CDF experiments at the Tevatron. The results from the most recent search will be discussed afterwards.

4.2.1.1 Search Algorithm Used by the H1 Experiment

The approach employed by the H1 experiment is a generic search for deviations from the SM prediction at large transverse momenta. Events containing electrons, muons, jets, photons or neutrinos are investigated. The analysis covers phase space regions where the SM prediction is sufficiently precise to detect anomalies without relying on assumptions about any new physics models that extend the SM. Multiple Monte Carlo event generators are combined to simulate events for all SM processes which have large cross sections or are expected to be dominant for particular final states. The different SM processes are generated with large statistics in comparison to the dataset used for the analysis, and the events are passed through a full detector simulation.

Final states containing at least two of the objects mentioned above with transverse momenta of greater than 20 GeV in the polar angle range of the detector between 10 to 140 degrees are investigated to search for deviations using the method described below.

First, the overall event yields of the final states are compared with the SM expectation. Then the distributions of the invariant mass M_{all} and of the scalar sum of transverse momenta $\sum P_T$ of the final state objects are investigated. New physics that may be visible as an excess or a deficit in one of these distributions is searched for. In a second step of the search, these distributions are systematically investigated using a dedicated algorithm which locates the region with the largest deviation of the data from the SM prediction. The search algorithm includes the calculation of the probability of occurrence of such a deviation, both for each event class individually, and also globally for all classes combined.

The search algorithm is discussed in further detail below, describing the four major steps involved.

- First, for the search based on the M_{all} and $\sum P_T$ distributions, the regions of search are defined for the kinematic distributions. A region in the M_{all} and $\sum P_T$ distributions is defined as a set of connected histogram bins with a size of at least twice the resolution. All possible such regions of any width and at any position in the histograms are considered for the search of deviations. The number of data events (N_{obs}), the SM expectation (N_{SM}) and its total systematic uncertainty (δN_{SM}) are calculated for each region. Figure 4.7, taken from [70], illustrates how the region of interest is identified.
- Second, to quantify the deviation in the data compared to the SM expectation in each region, a statistical estimator p is defined, as described in Eq. (4.6), derived from the convolution of the Poisson probability density function (pdf) to account for statistical errors with a Gaussian pdf to include the effect of systematic uncer-

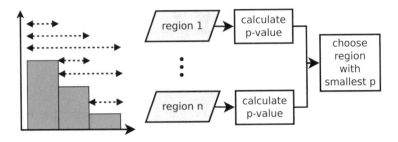

Fig. 4.7 Illustration for the selection of the region of interest [70]

tainties. The p value calculation also includes a normalisation factor that is given in Eq. (4.7).

$$p = \begin{cases} A \int_0^\infty db\, G(b; N_{SM}, \delta N_{SM}) \sum_{i=N_{obs}}^\infty \frac{e^{-b}b^i}{i!} & \text{if } N_{obs} \geq N_{SM} \\[3mm] A \int_0^\infty db\, G(b; N_{SM}, \delta N_{SM}) \sum_{i=0}^{N_{obs}} \frac{e^{-b}b^i}{i!} & \text{if } N_{obs} < N_{SM} \end{cases} \tag{4.6}$$

$$A = 1 \Big/ \left[\int_0^\infty db\, G(b; N_{SM}, \delta N_{SM}) \sum_{i=0}^\infty \frac{e^{-b}b^i}{i!} \right] \tag{4.7}$$

Note that both positive and negative deviations are considered.

The region of greatest deviation is the region having the smallest p-value, p_{min}. This defines which region is of most interest. Such a method is used to find narrow resonances and single outstanding events as well as signals spread over large regions of phase space in distributions.

- Third, the trial factor or the look-elsewhere effect is also accounted for. The probability that a fluctuation with a p-value at least as small as p_{min} occurs anywhere in the distribution is estimated using pseudo-experiments. Pseudo-experiments are created by filling many independent hypothetical data histograms with the number of events diced according to the pdfs of the SM expectation. For each such hypothetical data histogram, the algorithm is run to find the region of greatest deviation, and the corresponding p_{SM} is calculated. Then the value of $\hat{P}$ (called $\tilde{P}$ in other experiments) is defined as the fraction of hypothetical data histograms with a p_{SM} equal to or smaller than the p_{min} value obtained from the data. The thus obtained $\hat{P}$ is a measure of the statistical significance of the deviation observed in the data. For exclusive final states, the $\hat{P}$ values can be used to compare the results of different event classes. Depending on the final state, a p_{min}-value of $5.7 \cdot 10^{-7}$, that corresponds to the nominal '5σ' deviation that is considered to be significant, corresponds to a value of $-\log_{10} \hat{P}$ between 5 and 6.

- Finally, since this analysis would potentially cover a large number of final states, an estimate of the overall agreement of the observed data with the SM expectation is calculated. For this purpose, the probability of observing an event class with a given $\hat{P}$ value or a smaller one is calculated with MC experiments. MC experiments are defined as a set of hypothetical data histograms for the kinematic distributions considered following the SM expectation with an integrated luminosity corresponding to the dataset on which the search is performed. On repeating the entire search process using the MC experiments multiple times, the expectation for the $\hat{P}$ values observed in the data is then obtained from the distribution of corresponding $\hat{P}^{SM}$ values obtained from all MC experiments. The probability to find a $\hat{P}$ value smaller than the minimum observed in the data can thus be calculated to quantify the global significance of the observed deviation.

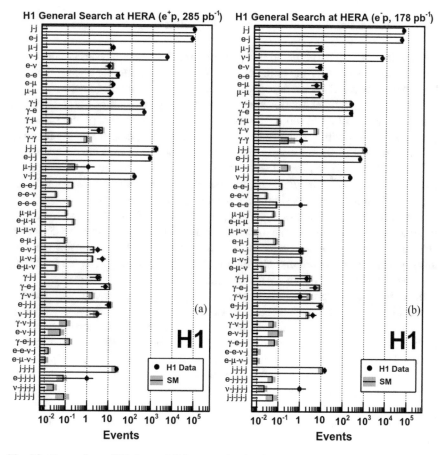

Fig. 4.8 Comparison of H1 data and SM expectation for all event classes with observed data [78]. Both statistical and systematic uncertainties are taken into account

Fig. 4.9 Distribution of the negative logarithm of $\hat{P}$ obtained by the H1 search in e^+p data, based on the invariant mass distributions of the event classes analysed. Note that the event class with two electrons (with all charge combinations) has a particular low $\hat{P}$ value. The distribution shown here is part of Fig. 6 in Ref. [78]

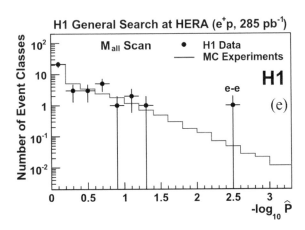

4.2.1.2   Results of the Search at the H1 Experiment

The results of the general model independent search for new physics conducted at the H1 experiment using the complete $e^{\pm}p$ collision data sample collected by the H1 experiment at HERA are described in Ref. [78] and summarised here.

The dataset investigated corresponds to the total integrated luminosity of $463\,\mathrm{pb}^{-1}$. While analysing all event topologies involving isolated electrons, photons, muons, neutrinos and jets with transverse momenta above $20\,\mathrm{GeV}$, data events are found in 27 different final states and events with up to five high p_T objects are observed.

The total event yields for H1 data and SM prediction are compared in Fig. 4.8, taken from Ref. [78]. The yields are compared for all event classes, separately for e^+p and e^-p initial states. A significant deviation is not seen.

For each final state analysed, deviations from the SM are searched for in the invariant mass and sum of transverse momenta distributions using the dedicated algorithm described in the previous section. In addition, the final states are also investigated in terms of angular distributions and energy sharing between final state particles, the details of which are not described here for the sake of brevity.

In general, good agreement is observed when comparing the data with the SM expectation in the phase space covered by this analysis. The largest deviation is found in the two-electron final state, at high invariant masses and corresponds to a probability of 0.0035. This can be seen in Fig. 4.9, which shows the $\hat{P}$ distribution obtained from the invariant mass distributions of all event classes, for the e^+p dataset.

The probability to observe an SM fluctuation with that significance or higher for at least one event class is estimated to be 12%. Hence, no claim for evidence of physics beyond the SM is made. The analysis shows that there has been good understanding of the high p_T physics phenomena in the context of the SM phenomena achieved by the H1 experiment at the HERA collider.

Chapter 5
General Model Independent Searches at the LHC

The Large Hadron Collider (LHC) at CERN is the most powerful particle accelerator in history and it continues to be in operation today. The LHC consists of a 27-km ring of superconducting magnets with a number of accelerating structures to accelerate the particles that are to be collided. The LHC was designed to collide protons with protons at a center-of-mass energy of 14 TeV. The LHC has already delivered proton–proton collision at center-of-mass energies of 7, 8 and 13 TeV since the beginning of full operations in 2010, and significantly large datasets have already been recorded by the LHC-based experiments. Separately, the LHC also collides heavy ions for the study of dense strongly interacting matter, with collisions of lead ions with lead ions or protons.

The major experiments based at the LHC are the A Toroidal LHC ApparatuS (ATLAS) experiment [67], the Compact Muon Solenoid (CMS) experiment [68], the Large Hadron Collider beauty (LHCb) experiment [69] and the A Large Ion Collider Experiment (ALICE) experiment. The ATLAS and CMS experiments consist of general-purpose particle detectors and have a large and diverse physics programme, ranging from the search and studies of the Higgs boson, performing precision standard model measurements and studies of strongly interacting matter at high densities to the search for new physics beyond the standard model. The LHCb experiment focuses primarily on the physics of b-hadrons. The ALICE experiment consists of a heavy-ion detector that is designed to study the physics of strongly interacting matter at extreme energy densities, where a phase of matter called quark–gluon plasma forms.

The main highlight of the physics programme of the LHC so far has been the discovery of the Higgs boson by the ATLAS and CMS experiments [5, 6]. The search for BSM physics phenomena is a major goal of the continuing physics programme of the LHC.

The LHC collides particles at the highest energy probed in collider experiments till date. The LHC produces such high energy collisions at high intensity, with the

© The Author(s), under exclusive license to Springer Nature Switzerland AG 2020
S. S. Ghosh et al., *General Model Independent Searches for Physics Beyond the Standard Model*, SpringerBriefs in Physics,
https://doi.org/10.1007/978-3-030-53783-8_5

intensity of the collisions expected to become even higher when the LHC is upgraded to the High-Luminosity LHC (HL-LHC) in the future. As such, the LHC has delivered large datasets at unprecedented energies and is expected to deliver more in the future. This presents the ideal opportunity to search for new physics phenomena, in the large datasets that provide access to regions of phase space that were not accessible before. Of the experiments at the LHC, the ATLAS and CMS experiments have a particularly broad programme consisting of several physics analyses that target the search for BSM physics. It is important that signs of new physics phenomena that might be hidden in the large datasets are not missed, and general model independent searches are ideal for such an endeavour.

General model independent searches have been conducted by both the ATLAS and the CMS experiments, and these are described in Sects. 5.1 and 5.2, respectively.

5.1 Model Independent Searches at the ATLAS Experiment

ATLAS is a general-purpose particle physics experiment at the LHC at CERN. The ATLAS detector is designed to be able to exploit the discovery potential and the huge range of physics opportunities that the LHC provides. The ATLAS detector is described in detail in Ref. [67].

The ATLAS experiment has a comprehensive physics programme, with the search for BSM physics being an important part. As such, the ATLAS experiment has several dedicated search analyses that target a wide range of diverse BSM physics models. The ATLAS experiment is also well suited to conduct a general model independent search on the LHC data.

The most recent general model independent search analyses at the ATLAS experiment is based on 3.2 fb^{-1} of proton collision data at a center-of-mass energy of 13 TeV collected during 2015 and is described in detail in Ref. [79]. The general search strategy used in the ATLAS general model independent search is described below in Sect. 5.1.1 and the results are summarised later in Sect. 5.1.2.

5.1.1 The ATLAS General Search Algorithm

The analysis strategy is built around the assumption that a signal of new physics can be identified as a statistically significant deviation of the event counts in the data from the expectation in a specific data selection region. The general model independent search relies on Monte Carlo (MC) simulations to estimate the SM expectation to compare with the observed data. The primary goal of the analysis is to identify selections for which the data deviates significantly from the SM expectation. These selection regions or final states can then be specifically chosen as data-derived signal regions to conduct a dedicated search analysis to determine the level of significance of deviation using another independent dataset. This would be the case for any dedicated search analysis with the difference being that for most such dedicated search analyses, the final state or data selection regions are chosen based on theoretical motivations corresponding to particular BSM model(s). This would have the advantage of a more reliable background expectation in the selected final states, which should allow an increase in signal sensitivity compared to a strategy that only relies on MC expectations with a typically conservative evaluation of uncertainties.

In the analysis, events are selected based on quality and trigger criteria and then classified according to the type and multiplicity of reconstructed objects into different final states or event classes. The reconstructed physics objects considered are electrons, muons, photons, jets, b-tagged jets and missing transverse momentum that pass selection criteria to be well-reconstructed objects with high transverse momentum (p_T). For the SM expectation, MC samples with sufficient statistics for all relevant background processes are included. While estimating the background model, the corresponding theoretical and experimental uncertainties are also included for the MC prediction used.

The complete search strategy is divided into seven steps that are described below.

- The first step is to select and classify events in data and from MC simulation. Events selected based on quality and trigger criteria are classified into final states or event classes based on the type and multiplicity of reconstructed objects that were mentioned above, that are electrons, muons, photons, jets, b-tagged jets and missing transverse momentum. Event classes (or channels) are then defined as the set of events with a given number of reconstructed objects for each type, for example, the event class containing two muons and a jet.

- The second step is to evaluate systematic uncertainties to perform validation studies. The modelling uncertainties in the background estimate arise from experimental effects, and the theoretical accuracy of the prediction of the cross section and acceptance of the MC simulation. These effects are evaluated for all contributing background processes as well as for benchmark signals.

  While the method for validation used for dedicated search analyses is to use control regions that are similar but orthogonal in selection to the probed signal region to validate, study or improve upon the MC-based background prediction, this approach becomes difficult for a general search in which a large number of final states are explored. Validation regions are used to test the validity of the background model prediction with data. To get around this difficulty and to verify the proper modelling of the SM background processes, several validation distributions are defined using inclusive selections for which observable signals for new physics are excluded based on past studies. If these validation distributions show problems in the MC modelling, either corrections to the MC backgrounds can be applied or the affected event classes can be excluded from the analysis.

- The third step is to identify sensitive variables and to define a statistical algorithm to probe for deviations between the data and the simulation expectation.

  Only a few observables are considered in order to avoid a large increase in the number of hypothesis tests. This helps in the computation process and also reduces the effect of the trial factor or look-elsewhere effect that would increase the rate of deviations due to background fluctuations if a large number of observables were to be considered. The observables considered are as follows:

- The effective mass m_{eff} (defined as the sum of the scalar transverse momenta of all objects plus the scalar missing transverse momentum),
- the total invariant mass m_{inv} (defined as the invariant mass of all visible objects in the event without considering the missing transverse momentum),
- the invariant mass of any combination of objects,
- event shape variables and more.

These variables are chosen since they have been widely used in searches for new physics and are sensitive to a large range of possible signals, manifesting either as bumps, deficits or wide excesses. The m_{eff} and m_{inv} distributions are taken in the form of histograms for each event class to scan for deviations of the data from the SM expectation.

To perform the scan, a statistical algorithm is used to scan the distributions described above in each event class, and to quantify the deviations of the data from the SM expectation. The statistical algorithm uses a test statistic for data selection regions, that is the p_0-values, which represents the probability of observing a fluctuation purely from the variation of the SM expectation, considering the involved uncertainties in the estimation of this expectation, that is at least as far from the SM expectation as the observed number of data events in a given region. This is the local p_0-value that must be corrected for trial factor. The algorithm identifies the data selection that has the largest deviation in the distribution of the investigated observable by testing many data selection regions to find the region with the smallest p_0-value.

The p_0-value test statistic used in this analysis is described in Eqs. (5.1)–(5.3). In the equations, n is the independent variable of the Poisson probability mass function (pmf), N_{obs} is the observed number of data events for a given selection, $P(n \leq N_{obs})$ is the probability of observing no more than the number of events observed in the data and $P(n \geq N_{obs})$ is the probability of observing at least the number of events observed in the data. N_{SM} represents the expectation for the number of events with its total uncertainty δN_{SM} for a given selection. The effects of systematic uncertainties and statistical uncertainties are taken into account with the convolution of the Poisson pmf (with mean x) with a Gaussian probability density function (pdf), $G(x; N_{SM}, \delta N_{SM})$ with mean N_{SM} and width δN_{SM}.

$$p_0 = 2 \cdot min\left[P(n \leq N_{obs}), P(n \geq N_{obs})\right] \tag{5.1}$$

$$P(n \leq N_{obs}) = \int_0^\infty dx\, G(x; N_{SM}, \delta N_{SM}) \sum_{n=0}^{N_{obs}} \frac{e^{-x} x^n}{n!} + \int_{-\infty}^0 dx\, G(x; N_{SM}, \delta N_{SM}) \tag{5.2}$$

$$P(n \geq N_{obs}) = \int_0^\infty dx\, G(x; N_{SM}, \delta N_{SM}) \sum_{n=N_{obs}}^{\infty} \frac{e^{-x} x^n}{n!} \tag{5.3}$$

For the scan of a distribution in an event class, the region that gives the smallest p_0-value is identified as the region with the largest deviation. The smallest p_0 for a given channel is defined as $p_{channel}$, corresponding to the local p_0-value of the largest deviation in that channel.

Certain regions are not considered for the scan if such regions have large uncertainties in the expectation arising due to a lack of MC events or from large systematic uncertainties. To ensure that this does not cause the overlooking of possible deviations, p_0-values of selection regions with more than three data events are monitored separately while single outstanding events with unusual object multiplicities can be seen as separate event classes.

- The fourth step is to account for the trial factor, which quantifies the look-elsewhere effect, through the generation of pseudo-experiments. This is done to take into account the probability that for a given observable one or more deviations of a certain size could occur somewhere from the SM expectation estimation in the event classes considered.

For the pseudo-experiments, pseudo-data distributions are produced from the SM expectation estimate while taking into account both statistical and systematic uncertainties by drawing pseudo-random data counts for the SM expectation distribution histogram bins from the convolved pmf described previously in Eqs. (5.2) and (5.3). Each pseudo-experiment consists of the same event classes as those considered when applying the search algorithm to data, with the data counts being replaced by pseudo-data counts which are generated from the SM expectation as described above. The effect of correlations between bins of the same distribution or between distributions of different event classes are also taken into account when generating pseudo-data for pseudo-experiments; however, the correlations between distributions of different observables are not taken into account, since the results obtained for different observables are gained separately during the statistical interpretation.

The search algorithm is then applied to each of the distributions, resulting in a $p_{channel}$-value for each event class for the pseudo-experiment rounds. The $p_{channel}$ distributions of many pseudo-experiments and their statistical properties are compared with the $p_{channel}$ distribution obtained from data, in order to interpret the test statistics in a frequentist manner, with the value of the fraction of pseudo-experiments having a $p_{channel}$-value smaller than a given value p_{min}, indicating the probability of observing such a deviation by chance, taking into account the number of selection regions and event classes tested. So, comparing with the value obtained from the data scan, the probability for the observed deviation arising by chance can be estimated. This can then be interpreted in the form of the statistic $P_{exp,i}(p_{example})$ that represents the probability of observing, without the presence of any signal in the data, at least i number of $p_{channel}$-value(s) ($i = 1$ for the probability of observing at least one such value) with a $p_{channel}$-value of less than a particular value, $p_{example}$ in the above description that could refer to a particular value observed from the data scan.

Specific thresholds are set so that deviations observed that show a greater deviation than a predefined threshold are to be probed for signs of new physics phenomena.

The observation of one, two or three $p_{channel}$-values in data below the corresponding p_{min} threshold, set corresponding to an observation with a $P_{exp,i} < 0.05$, is required for the selection regions that yielded these deviations to be considered as signal regions that can be tested for signs of new physics phenomena, using the method outlined below in the seventh step.

- The fifth step, before coming to the results of the scans, is to test the sensitivity of this procedure.

  The sensitivity is evaluated with two different methods, one uses a modified background estimation through the removal of SM processes, the other adds a signal contribution to the pseudo-data sample.

  In the first method, a rare SM process is removed from the background model. The search algorithm is applied again to test the data or pseudo-experiments generated from the unmodified SM expectation, against the modified background. This was done for the SM WZ process and the $t\bar{t}\gamma$ process. For the tests with each process, $P_{exp,i} < 0.05$ is found for all three cases ($i = 1, 2, 3$), thus confirming the sensitivity of this method towards detecting such processes.

  In the second method, pseudo-experiments are used to test the sensitivity of the analysis to benchmark signal models of new physics, with the model with a Z' boson based on the sequential extension of the SM gauge group (SSM) considered [80]. The prediction of a model is added to the SM prediction, and this modified expectation is used to generate pseudo-experiments. The search algorithm is applied to the pseudo-experiments. The results are found to verify the sensitivity of this method towards detecting such processes.

- The sixth step is to perform the above-described scans with respect to the data distributions and obtain the results.

  As mentioned before, finding one or more deviations in the data with $P_{exp,i} < 5\%$ triggers a dedicated analysis that uses the data selection in which the deviation is observed as a signal region.

  However, even if no significant deviations are found, the outcome of the analysis technique includes important information such as the number of events and expectation per event class, a comparison of the data with the SM expectation in the distributions of observables considered, the scan results and the comparison with the expectation from pseudo-experiments.

- The seventh and final step of the analysis is only performed if one or more deviations in the data with $P_{exp,i} < 5\%$ are found. Then a dedicated analysis is to be performed specific to the selection region(s) where the deviation is found, in order to probe the deviation in terms of possible signatures of new physics phenomena. The dedicated analysis can be performed on the original dataset where the deviations were investigated or on an independent dataset. For the analysis on the original dataset, the background prediction can be improved using control regions to control and validate the background modelling. The dependence on the simulation to estimate the background would be reduced and this method would better constrain the background expectation and uncertainty. A dedicated analysis might show an insignificant deviation, in which case it can be inferred that the deviation seen before was due to mismodelling or incomplete estimation of the background.

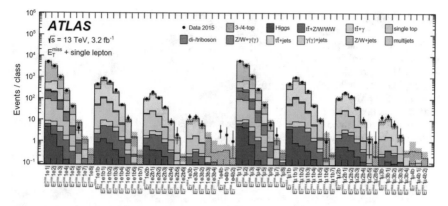

Fig. 5.1 The experimental and simulated event yields, for a subset of event classes analysed by ATLAS [79]. Only those with missing transverse energy, at least one lepton and at least one jet are displayed

If the dedicated analysis does confirm the deviation observed, the data selection region in which the deviation is observed defines a data-derived signal region that is to be tested in an independent new dataset with a similar or larger integrated luminosity. For this statistically independent analysis, since the signal region is known, the corresponding data can be excluded from the analysis until the very end, in a so-called 'blind' analysis, to minimise any possible bias in the analysis. This would also exclude the need to take into account the look-elsewhere effect. At this stage, the deviation can also be interpreted in terms of different known BSM models.

5.1.2 The ATLAS General Search Results

The general model independent search analysis described in the previous section was implemented at the ATLAS experiment using the 3.2 fb^{-1} dataset collected during proton collisions at a center-of-mass energy of 13 TeV in 2015 [79]. In this dataset, exclusive event classes containing electrons, muons, photons, b-tagged jets, non-b-tagged jets and missing transverse momentum have been scanned for deviations from the MC-based SM prediction in the distributions of the effective mass and the invariant mass.

The analysis of the dataset showed no significant deviations and consequently no dedicated analysis using data-derived signal regions has been performed.

The event yields for a subset of 60 out of a total of 704 event classes are compared to the SM prediction in Fig. 5.1, taken from Ref. [79]. Anomalies are not found in this comparison, nor in the analysis of the event yields in the other 644 classes.

A set of scans for the different final states including a comparison of the total event yields has been presented. These show a good agreement between the ATLAS

Fig. 5.2 Example distribution for events with six jets [79]. The kinematic variable m_{eff} is explained in the text. The region of interest, ROI, near 4000 GeV, is marked. The SM expectation in this region is 10.01 ± 3.05, while three events have been measured

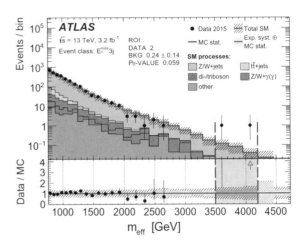

data and simulation in a wide range of diverse final states, thus demonstrating the understanding of the SM processes and the ability to simulate them while including the detector effects in a range of diverse event topologies.

An example for a distribution with a negative deviation, which is not statistically significant is shown in Fig. 5.2, taken from Fig. 4 in Ref. [79].

The general model independent search strategy discussed here will be useful to search for signals of unknown new physics phenomena in the subsequent LHC Run 2 datasets collected by the ATLAS experiment.

5.2   Model Independent Searches at the CMS Experiment

The Compact Muon Solenoid (CMS) experiment consists of a general-purpose detector at the CERN LHC. While similar to ATLAS in certain aspects, CMS is more compact and uses a different detector concept, including a different design for the magnet system. The CMS detector is described in Ref. [68].

The CMS experiment has a broad physics programme ranging from studying the Standard Model (including the Higgs boson) to searching for BSM physics phenomena. CMS has a large programme of dedicated searches for multiple BSM physics models. Like the ATLAS experiment, CMS is also well suited to conduct a general model independent search on the LHC data.

MUSiC or the Model Unspecific Search in CMS analysis is the general model independent search that is conducted at the CMS experiment. The idea goes back to the L3 experiment [81]. The public reports of the CMS MUSiC analysis can be found in Refs. [82–84].

5.2.1 MUSiC: Model Unspecific Search in CMS

In line with the general model independent search approach, the MUSiC search analysis aims to be unbiased by models of physics beyond the Standard Model and to be as inclusive as possible while searching for significant deviations in the CMS data compared to the SM expectation, which could be signs of new physics. The analysis strategy is designed to use an automated approach that is valid over the entire range of final states and kinematic distributions that are considered for the search.

The modelling of the known SM background processes is based solely on Monte Carlo simulation. The version of the MUSiC algorithm discussed here took the following physics objects into account: electrons, muons, photons, jets, bottom-jets and the Missing Transverse Energy (MET) of the event. The analysis strategy consists of the following steps: (1) the classification of events into different event classes, (2) the definition of kinematic distributions of interest, followed by (3) the scanning procedure to search for deviations and (4) accounting for the look-elsewhere effect using pseudo-data generation and finally (5) interpreting the search in terms of a global overview of the scan results. These steps are described in detail below.

1. The first step of the MUSiC analysis is the selection and classification of events. Broadly, lepton (electron or muon) triggered events are selected from the experimental data and simulation of the SM expectation. They are classified into different event classes, or final states, while taking into account the physics objects that are present in the event, considering electrons, muons, photons, jets, bottom-jets and the Missing Transverse Energy, MET. Specific selection criteria are set to identify well-reconstructed and isolated objects with sufficient transverse momentum. Event classes must contain at least one lepton. Each event is sorted into three different types of event classes:

   - Exclusive event classes for events where all events are reconstructed with exactly the selected objects.
   - Inclusive event classes that contain events that include the set of selected objects but may contain additional objects.
   - Jet-inclusive event classes, which are similar to inclusive classes but restrict additional allowed objects to jets.

   For an illustrative example, an event containing one electron, one muon and a jet : $1e + 1\mu + 1$ jet would be classified into the $1e + 1\mu + 1$ jet exclusive event class only, but for the inclusive classes, the event would contribute to the $1e + 1\mu + 1$ jet inclusive event class and also the $1e$ inclusive event class, 1μ inclusive event class, $1e + 1\mu$ inclusive event class, $1e + 1$ jet inclusive event class and the $1\mu + 1$ jet inclusive event class. For the jet-inclusive classes, this event would contribute to the $1e + 1\mu$ jet-inclusive event class and the $1e + 1\mu + 1$ jet jet-inclusive event class.

2. For the second step, three kinematic variables are chosen to be probed in the MUSiC analysis. This choice is based on their promise in terms of sensitivity

to new physics phenomena, and that they are well defined in the different final states. These three are as follows:

- $\sum |\vec{p}_T|$ or S_T, that is the scalar sum of the transverse momenta of all the physics objects which are included in the definition of the event class. This observable is sensitive to high energy phenomena that might appear in the tails of such distributions. For example, in the exclusive event class $1e + 1\mu + \text{MET}$ the transverse momenta of electron, muon and also MET are summed up. For the inclusive class $1e + 1\mu + \text{MET} + X$ the transverse momentum associated with X is not included in the calculation of S_T.
- The invariant mass M, that is the combined invariant mass of all the physics objects considered in the event class. The transverse mass is used for event classes that have missing transverse energy. This observable is sensitive to resonances associated with BSM particles.
- The Missing Transverse Energy (MET) in the event classes with significant MET. The MET is sensitive to heavy or highly boosted weakly interacting particles associated with some BSM models.

The above-mentioned distributions, in the form of binned histograms, are then scanned for deviations between CMS data and simulated SM expectation in the different event classes.

3. The third step of the MUSiC algorithm is to perform scans to identify deviations between the measured data and the simulated SM expectation. There are two important aspects to the statistical analysis to scan for deviations, one is to have a test statistic to assess the deviations quantitatively, and the second is to define regions within the kinematic distributions where a quantitative comparison is performed.

A p-value statistic is used to quantify observed deviations. The p-value statistic used is described in Eq. (5.4), where N_{data} is the number of data events observed, N_{SM} is the number of expected events from the simulation of the SM expectation, σ_{SM} denotes the uncertainty on the number of expected events, combining the statistical and systematic uncertainties and C is the normalisation factor. The p-value statistic described here is a combined Frequentist–Bayesian probability value, calculated from Poisson statistics convoluted with a Gaussian prior function to include the uncertainties.

$$
p = \begin{cases} \displaystyle\sum_{i=N_{data}}^{\infty} C \cdot \int_0^{\infty} d\lambda \, e^{-\frac{(\lambda - N_{SM})^2}{2\sigma_{SM}^2}} \cdot \frac{e^{-\lambda}\lambda^i}{i!} & \text{if } N_{data} \geq N_{SM} \\[20pt] \displaystyle\sum_{i=0}^{N_{data}} C \cdot \int_0^{\infty} d\lambda \, e^{-\frac{(\lambda - N_{SM})^2}{2\sigma_{SM}^2}} \cdot \frac{e^{-\lambda}\lambda^i}{i!} & \text{if } N_{data} < N_{SM} \end{cases}
\tag{5.4}
$$

Note that both excesses and deficits are looked for. The deviations to be quantified by the p-value described above could be narrow peaks affecting a few histogram bins, or they could be wider deviations. In order to identify both these different types of deviations, the MUSiC algorithms construct regions from connected bins and perform the comparison of CMS data and simulation by calculating the p-value in the different regions. A 'region' is defined as any contiguous combination of bins. All bins in the histograms of the considered distributions are successively combined to regions by adding up their individual contributions, and a p-value is calculated. It is taken care that each region has a reliable estimate of the SM expectation by vetoing regions where specific criteria associated with reliable background modelling are not met. The region with the smallest p-value is then selected as the region with the largest deviations and referred to as the Region of Interest (RoI). The scan is performed and the p-value for the region of interest is calculated for each of the kinematic distributions scanned in each of the event classes considered. The p-value thus obtained must be corrected for the look-elsewhere effect, and the procedure to do so is described next.

4. In the fourth step, the post-trial probability is estimated to correct the p-value to take into account the look-elsewhere effect by using pseudo-experiment rounds. It is important to perform this correction since it accounts for the increased probability to observe a deviation arising from the fluctuation in the background, if a large number of regions is considered.

For the pseudo-experiments, pseudo-data are generated for a single class distribution by generating randomised values for each bin, based on the simulated SM expectation. This is done by varying the yields based on the uncertainties in the expectation to create an ensemble of expected values that are based on the expected values given the SM only hypothesis.

A large number of rounds of pseudo-experiments are thus created and every round of a pseudo-experiment is scanned for a RoI and the local p-value is calculated. The number of trials resulting in a local p-value (p_{min}) smaller or equal to the one found in the data to simulation comparison (p_{data}) is determined ($N_{pseudo}(p_{min} < p_{data})$) and divided by the full number of trials (N_{pseudo}):

$$\tilde{p} = \frac{N_{pseudo}(p_{min} < p_{data})}{N_{pseudo}} \tag{5.5}$$

This fraction is the post-trial p-value ($\tilde{p}$-value), representing a statistical estimate of how probable it is to see a deviation at least as strong as the observed one in any region of the distribution.

5. In the final step, a global overview of the scan results is presented for each kinematic distribution probed for the different event class types (exclusive, inclusive and jet-inclusive). It shows the distribution of the number of event classes in bins of the quantification of the observed deviations ($-\log_{10}[\tilde{p}]$). The observed distribution is compared with the expected distributions that are calculated based on the pseudo-experiments, thus giving an expected distribution of the deviations along with uncertainty bands corresponding to the 1σ and 2σ bands. If there are observed deviations that cannot be accounted for based on the expectations, then this is a sign for new physics phenomena.

This method of comparison has the benefit of showing if there are a number of smaller deviations observed, along with the case where there might be one or a few event classes showing a large deviation. Some BSM signals could show smaller deviations spread out over several different final states instead of a large deviation in certain final states as is the case for other models. This method would be able to detect both such types of deviations.

Having described the search algorithm, it is important to mention that the sensitivity of the MUSiC search algorithm has also been tested. Similar to the studies performed for the ATLAS general search, the sensitivity of the MUSiC algorithm has also been tested using the two methods of testing the sensitivity to a benchmark BSM signal model added on top of the background using pseudo-data and also by testing the sensitivity of the MUSiC algorithm to detect a rare SM process after removing the SM process from the background expectation. The MUSiC algorithm was found to be sensitive in both such tests.

5.2.2 Results of the MUSiC Analysis with a CMS 13 TeV Dataset

The most recent public result of the MUSiC analysis is based on the CMS dataset collected during proton collisions at a center-of-mass energy of 13 TeV during 2016 corresponding to an integrated luminosity of 35.9 fb^{-1} [84]. These are preliminary results with the final set of results still awaiting publication. Events selected with at least one lepton are used.

The classification algorithm described in the previous section resulted in a total of 498 exclusive, 571 inclusive and 530 jet-inclusive event classes with at least one data event. For the number of classes in MC a lower threshold of 0.1 event is required. The step of comparing the event yields in the CMS data and the SM simulation prediction shows no significant deviation large enough to be investigated as a hint of new physics. Comparison of the event yields in a range of different final states between the recorded experimental data and the simulation of the SM is also presented in Ref. [84].

For all the event classes the three kinematic variables described above, S_T (scalar sum of transverse momenta), M (invariant mass or transverse mass M_T) and MET are studied, where applicable.

To test the sensitivity of the MUSiC approach, a Monte Carlo-based study is performed. In particular, the prediction of a BSM model [85] with a new heavy charged vector boson W$'$ with a mass of 3 TeV is added to the Standard Model background, see Fig. 5.3. The statistics corresponds to the 13 TeV dataset recorded in the year 2016. Such a boson could, similarly to the Standard Model W boson, decay into muon plus (undetected) neutrino. The simulated transverse mass M_T distribution for the event class $1\,\mu + $ MET is clearly sensitive to this new particle, producing a deviation well beyond the expectation from an SM only hypothesis.

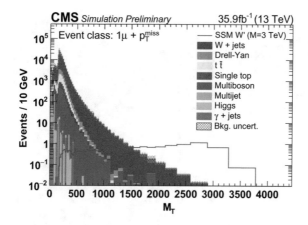

Fig. 5.3 Distribution of the transverse mass for events with one muon plus missing transverse momentum MET $\equiv p_T^{miss}$, measured by CMS [84]. The transverse mass is calculated from the momentum components perpendicular to the beam axis, since the longitudinal component of the missing momentum is unknown. A hypothetical signal arising from the decay of a heavy W′ boson with a mass of 3 TeV is also shown, compared to the simulation of the SM processes. The W′ signal is clearly not compatible with the SM

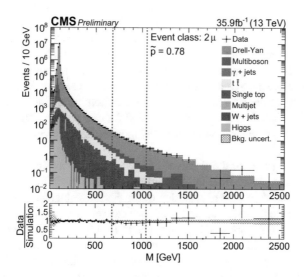

Fig. 5.4 Distribution of the invariant mass for the exclusive two muon event class, measured by CMS [84]. For the standard model prediction the various contributions from top pair production, the Drell–Yan process, etc. are shown separately. The search window with the biggest (but insignificant) discrepancy found by the MUSiC search algorithm, the region of interest RoI, is marked by the two dashed vertical lines. Overall there is good agreement between theoretical prediction and experimental data, as is quantified by the $\tilde{p}$ value of 0.78

Fig. 5.5 Distribution of $\tilde{p}$
values, measuring the
significance of differences
between CMS data and SM
prediction [84]. The $\tilde{p}$ values
are determined from the
invariant mass M
distributions of all exclusive
event classes

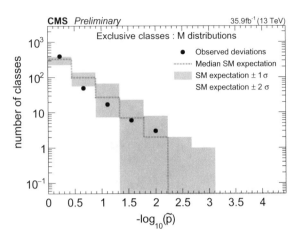

The MUSiC Region of Interest scan is performed comparing the SM simulation
to CMS data for exclusive, jet-inclusive and inclusive classes. While there are small
discrepancies, no statistically significant deviations show up, and no evidence for
new physics is found. An example is shown in Fig. 5.4. The measured distribution
of the invariant mass of the two muons in the 2μ event class is well described by
the Monte Carlo simulation. The peak near $M = 91\,\text{GeV}$ is due to the Z resonance.
Similarly heavier resonances would show up as peaks at higher mass values. The
MUSiC algorithm has found the biggest discrepancy in the region 680–1040 GeV,
this is the RoI. Here the Standard Model prediction is a bit above the measured event
yield, as can be seen best in the ratio plot at the bottom of the figure. The statistical
significance is small, thus there is no indication for new physics. This is an example
of a less significant deviation, and final states showing larger deviations are discussed
in Ref. [84].

The global overview of the $\tilde{p}$ values compares the distribution of the deviations
observed to that expected from the SM only hypothesis given the associated uncertainties. Figure 5.5 shows the distribution of the $\tilde{p}$ values obtained from all M distributions of the exclusive classes. First of all, one can see that very low $\tilde{p}$ values are not
found at all, since there are no data points above $\tilde{p} \approx 6 \cdot 10^{-3}$. Furthermore, the SM
prediction for this distribution, as determined from pseudo-experiments, reproduces
the measurements within uncertainties. Thus there is no hint for new physics. The $\tilde{p}$
distributions for the other kinematic variables, and for the inclusive classes, show a
similar picture.

Chapter 6
Comparison of Model Independent Searches

In the previous chapters, the model independent methods applied at collider experiments (CDF, D0; H1; ATLAS, CMS) were presented. Now we compare and evaluate these efforts.

The presented searches have several basic features in common. Some of the similarities can be attributed to the fact that more recent analyses profit from the experiences gained in earlier studies. Other common traits are plain necessities, like using automatized procedures or taking into account the look-elsewhere effect. Here follows a list of the common aspects:

- **BSM dependence**
  The methods do not refer to any specific BSM physics hypothesis, and to that effect the searches are model independent, indeed.
- **Standard Model prediction**
  The measurements are compared to Monte Carlo simulations of Standard Model processes, including parton showering, hadronization and detector response. One should keep in mind that those programmes have been to some extent verified and tuned using experimental data—under the assumption of absence of new physics. For example, some of the parameters of the PYTHIA generator [86, 87] are obtained in this way [88]. Thus there could be a small bias.
- **Dataset**
  The datasets used are large, ideally all recorded collider events are analysed. Excluded are final states for which the simulation is known to have shortcomings, for example, if higher order QCD processes are involved, or if the detector is reaching its limit in terms of triggering, particle identification or momentum measurement. Another exclusion criterium can be low statistics in the experimental data and/or in the Monte Carlo event sample, so that at best the yield can be measured and compared, but not differential cross sections.

© The Author(s), under exclusive license to Springer Nature Switzerland AG 2020   57
S. S. Ghosh et al., *General Model Independent Searches for Physics Beyond
the Standard Model*, SpringerBriefs in Physics,
https://doi.org/10.1007/978-3-030-53783-8_6

- **Classification**
  The datasets are structured in terms of event classes, defined by the particles identified in the final states, in particular electrons, muons, photons, jets and also missing transverse energy.
- **Observables**
  The experiments have looked both at the integrated cross section, that is the total event yield, and a couple of differential cross sections, for example, the scalar sum of the transverse momenta of all objects in one event. All published analyses have restricted the number of distributions analysed, in order to limit the look-elsewhere effect. For the same reason only one-dimensional distributions are considered. The Monte Carlo SM predictions are always normalised to the measured integrated luminosity.
- **Uncertainties**
  Statistical uncertainties are taken into account, in the form of Poisson statistics. Also systematic uncertainties and correlations are included, mostly by a Gaussian smearing of the SM prediction. Those can affect the shape and/or normalisation of distributions.
- **Deviations and Significance**
  With the exception of VISTA, deviations in a distribution are searched for by a sliding window of variable size. The biggest discrepancy is quantified by the local p-value. A post-trial p-value, here called $\tilde{p}$-value,[1] is obtained from a large number of pseudo-experiments. The $\tilde{p}$-values are a measure of the significance of the deviation found in a given distribution of a certain class.
- **Global comparison**
  The global agreement between experimental data and SM Monte Carlo prediction is obtained by a comparison of the corresponding distributions of the $\tilde{p}$ values obtained from all classes. This method allows to reveal new physics if it affects many classes, even if in individual classes the deviations are small.
- **Implementation**
  The analyses are automatized. This is a pure necessity, given that thousands of distributions need to be scrutinised. Finally, event classes with deviations between SM and measurement are inspected by experienced physicists, to judge if misreconstructions or other effects which are not properly modelled can be the cause.

While the common features prevail, there are quite a few differences in the details of the models employed. In the following, a non-exhaustive list is given:

- **Standard Model prediction**
  The VISTA algorithm does not fully rely on the SM Monte Carlo programmes. Therefore, in a first iteration of the analysis, theoretical and experimental correction factors are calculated by globally minimising the difference between many experimental and predicted distributions. This global approach is believed to minimise a possible bias.

---

[1] Some experiments use other notations, for example, $\hat{p}$ in H1.

- **Dataset**

  The selection cuts for the final state particles vary from experiment to experiment. CMS has excluded events with only jets in the final state, to limit the impact of QCD uncertainties. The kinematic cuts imposed depend also on the type of collider; for example, H1 includes a transverse momentum cut of 20 GeV and selects only particles in the polar angle range between 10° and 140°. The SLEUTH algorithm restricts the search to tails of kinematic distributions corresponding to high transverse momenta. This is based on the expectation that new physics will show up most likely at high mass scales, else it would have been found already. While here some model dependence comes in, the SLEUTH method leads to an analysis which is similar to the other approaches, and it investigates fewer regions. Therefore it profits from a smaller look-elsewhere effect.

- **Deviations and Significance**

  In VISTA the agreement between measurement and prediction is quantified by using a Kolmogorov–Smirnov test. In SLEUTH the window method is applied, but only the lower bound is varied, the higher one is set to infinity. Differences between measurement and SM predictions can be positive or negative. Only SLEUTH searches explicitly for an excess in the experimental data. The formulae for the p-value calculation for a given region in a kinematical distribution, see Eqs. (4.6) and (5.2)–(5.4), are nearly identical. The additional term in (5.2) suppresses negative MC event predictions.

- **Observables**

  Here all the experiments use a different number of kinematical observables and different quantities. However, most of them aim at new physics appearing at a high mass scale. ATLAS has only two observables (m_{eff} and m_{inv}, as defined in Sect. 5.1.1), while VISTA investigates several quantities, including angular observables.

- **Classification**

  In some analyses, bottom-jets are treated as separate object types. The charges of the leptons are taken into account in some analyses, see for example Fig. 4.6. Some simplifications are made in the event class definitions for the SLEUTH algorithm [76], this includes grouping of jets and of electron and muon events. In CMS both exclusive event classes like (exactly) $2\,e$ as well as inclusive classes like $2\,e + X$ are analysed.

In summary, the presented general model independent searches have reached a high level of maturity. There are differences between the methods used by the different experimental collaborations, some of them are simply due to different collider and detector types. The variation in methodology underlines that new ideas are being tried out. Overall, the performances of these searches are similar, and the comparison does not reveal a clear 'winner'.

Chapter 7
Outlook for General Model Independent Searches

Looking forward, there are exciting opportunities to perform general model independent searches for new physics phenomena and improve upon the existing methods employed for such searches by including several recent developments, some of which are discussed here. Progress on anomaly detection methods using machine learning techniques can contribute significantly to the development of general model independent search approaches going forward.

Several different implementations of the general model independent search approach have been described in the previous chapters. While following a similar broad template, each of the implementations has their unique features. These different implementations have been successful in comparing the data to the Standard Model expectation in a wide range of diverse event topologies and phase space regions. This provides confidence in pursuing with the general model independent search approach going forward.

Recent and forthcoming developments in different aspects relevant to collider-based high energy physics experiments can be used to enhance the general model independent searches in the future. Simulations of the different SM processes that include higher order theoretical calculations are becoming more readily available. Such improved estimations from simulation of the SM processes that constitute backgrounds for the general searches lead to the reduction in the uncertainty in the background estimation, thus making it possible to observe smaller deviations between experimental data and SM expectation. Improved understanding and modelling of the detector effects using improved tools such as using machine learning techniques would provide a closer comparison between what is observed in the data and what is simulated in terms of the detector effects. Such improved modelling of the behaviour of the expected background would reduce the reliance on specific data-driven estimation of the background that is employed in dedicated searches and cannot easily be used in general search analyses.

© The Author(s), under exclusive license to Springer Nature Switzerland AG 2020
S. S. Ghosh et al., *General Model Independent Searches for Physics Beyond the Standard Model*, SpringerBriefs in Physics,
https://doi.org/10.1007/978-3-030-53783-8_7

Furthermore, newly developed techniques to reconstruct more complex physics objects, such as top quarks or hadronically decaying W or Z bosons with high momentum, could be incorporated to expand the scope of the general model independent searches. In the implementations described in the previous chapters, general searches are limited to using more basic and comparatively easy to reconstruct physics objects such as electrons, muons, photons, hadronic jets, b-quark associated jets and missing transverse momentum. The inclusion of more complex reconstruction of other physics objects would allow general searches to probe regions of the phase space of different BSM models that were hitherto unexplored or for which only dedicated analyses have been sensitive to so far. Progress in available computing resources over time would also benefit general model independent search analyses, which often require intensive computing resources to perform the complex analysis of large amounts of data.

The general model independent search approaches described in the previous chapters largely rely on simulation to estimate the expected background, and search for deviations in the observed data compared to the expected background. There are also other methods that have been developed to search for new physics phenomena without relying on the inputs of specific BSM models. One such technique searches for 'bumps' or resonance like structures within an otherwise smooth data distribution. This is the so-called Bump Hunter approach, and more information about this approach can be found in Ref. [89]. This approach does not rely on simulation to model the background expectation, and the search algorithm relies only on the distributions observed in the experimental data to search for unexpected resonance like structures. This approach would be limited to such possible BSM scenarios where the signatures appear in the form of resonances in the distributions probed.

Expanding upon such an approach, newer techniques have been developed using modern machine learning tools. One example of this is the technique developed for anomaly detection for resonant new physics with machine learning that is described in Ref. [90]. This method uses classifiers trained directly on data to search for resonant like structures in the distribution of a selected kinematic observable that would appear as anomalies on top of a smoothly varying background, without any explicit input of any BSM physics model. Another proposed technique, described in Ref. [91], relies on adversarial neural networks, again trained directly on the data, to perform unsupervised searches for new physics that would appear as anomalies in the data. Such recent developments showcase the potential of using modern machine learning techniques to perform general model independent searches for new physics. An example of an experimental analysis using machine learning techniques to conduct model independent search for new physics phenomena can be found in Ref. [92] that describes a search for resonances in dijet final states conducted by the ATLAS experiment, yet this analysis is limited to the dijet final states and is not general in terms of covering a diverse range of final states.

While there are several ways to improve and enhance the scope of general model independent searches, some of which have been described above, another promising development is the availability of large datasets on which such searches can be performed, with the large datasets collected by the experiments at the LHC. In the coming years, the LHC is expected to be upgraded to the High-Luminosity LHC (HL-LHC) that would deliver even more data. A total of about 3000 fb^{-1} of proton collision data per experiment at a center-of-mass energy of 14 TeV is expected to be delivered by the HL-LHC by the end of its operation. This is an improvement by two orders of magnitude compared to the statistics used in the published model independent searches at the LHC. This would provide an excellent opportunity to conduct general model independent searches, particularly in order to not miss any signs of new physics phenomena that might be detectable in such large datasets.

There exists substantial theoretical motivation for physics beyond the Standard Model of particle physics, as has been described before. Despite extensive experimental searches, no direct indication of new physics beyond the Standard Model has been found so far. In the absence of any significant evidence of BSM physics, it is essential to have a broad search programme that would also be able to probe possible unexpected new physics scenarios. As such, it is imperative to continue to perform general model independent searches for new physics phenomena.

About the Authors

Saranya Samik Ghosh is a scientific researcher at RWTH Aachen University, working on the CMS experiment, on model independent search for new physics. He has been a member of the CMS Collaboration since 2011 and has worked on precision standard model measurements and studies of the Higgs boson in the past, having also contributed to the first experimental observation of the production of the Higgs boson in association with a pair of top quarks. He obtained his Ph.D. from the Tata Institute of Fundamental Research, Mumbai in 2015 and had worked at CEA Paris-Saclay before joining RWTH Aachen University. He is also actively involved in working on the muon detector system and on the object reconstruction efforts at the CMS experiment along with his work on physics analyses.

Thomas Hebbeker is professor for experimental physics at RWTH Aachen University in Germany. He has been working on the CMS experiment at CERN for more than twenty years, contributing to muon detectors and to data analyses. His team is in particular searching for new particles, such as extra W bosons, excited leptons, or supersymmetric particles. Another important focus is model independent search for new physics. He is also member of the Pierre Auger Collaboration, measuring ultra high energy cosmic rays.

Arnd Meyer, at RWTH Aachen University since 2003, has been searching for physics beyond the standard model for much of his scientific career. Having obtained his Ph.D. in physics from Hamburg University in the H1 experiment at DESY, he crossed the Atlantic to join first the CDF and then the D0 experiments at Fermilab, where he took on leadership positions in running the experiments. Since the early 2000s he has pioneered a number of BSM searches, including excited leptons at the Tevatron, unparticles and lepton flavour violation at the LHC, and model independent searches at both. He has been a member of the CMS Collaboration since 2006.

S. S. Ghosh et al., *General Model Independent Searches for Physics Beyond the Standard Model*, SpringerBriefs in Physics, https://doi.org/10.1007/978-3-030-53783-8

Tobias Pook has been a member of the CMS experiment at CERN since 2011 and worked on searches for extra spatial dimensions before he joined the research group for model independent searches as the lead analyst for the study of 13 TeV data. He developed and maintained major parts of the software framework for the model independent searches and helped to maintain the required GRID computing infrastructure.

References

1. F. Englert, R. Brout, Broken symmetry and the mass of gauge vector mesons. Phys. Rev. Lett. **13**, 321 (1964)
2. P.W. Higgs, Broken symmetries, massless particles and gauge fields. Phys. Rev. Lett. **12**, 132 (1964)
3. P.W. Higgs, Global conservation laws and massless particles. Phys. Rev. Lett. **13**, 508 (1964)
4. C.R. Hagen, G.S. Guralnik, T.W.B. Kibble, Global conservation laws and massless particles. Phys. Rev. Lett. **13**(585) (1964)
5. ATLAS Collaboration, Observation of a new particle in the search for the standard model Higgs boson with the ATLAS detector at the LHC. Phys. Lett. B **716**, 1 (2012)
6. C.M.S. Collaboration, Observation of a new boson at a mass of 125 GeV with the CMS experiment at the LHC. Phys. Lett. B **716**, 30 (2012)
7. https://texample.net/tikz/examples/model-physics/
8. F.J. Hasert et al., Search for elastic ν_μ electron scattering. Phys. Lett. B **46**, 121–124 (1973) [5.11 (1973)]
9. F.J. Hasert et al., Observation of neutrino like interactions without muon or electron in the Gargamelle neutrino experiment. Phys. Lett. B **46**, 138–140 (1973) [5.15 (1973)]
10. F.J. Hasert et al., Observation of neutrino like interactions without muon or electron in the Gargamelle neutrino experiment. Nucl. Phys. B **73**, 1–22 (1974)
11. G. Arnison et al., Experimental observation of lepton pairs of invariant mass around 95-GeV/c**2 at the CERN SPS collider. Phys. Lett. B **126**, 398–410 (1983) [7.55 (1983)]
12. M. Banner et al., Observation of single isolated electrons of high transverse momentum in events with missing transverse energy at the CERN anti-p p collider. Phys. Lett. B **122**, 476–485 (1983) [7.45 (1983)]
13. P. Bagnaia et al., Evidence for Z0 → e+ e- at the CERN anti-p p collider. Phys. Lett. B **129**, 130–140 (1983) [7.69 (1983)]
14. Planck Collaboration, Planck 2015 results. A&A **594**, 1 (2016)
15. Y. Fukuda et al., Evidence for oscillation of atmospheric neutrinos. Phys. Rev. Lett. **81**, 1562–1567 (1998)
16. Q.R. Ahmad et al., Measurement of the rate of $\nu_e + d \to p + p + e^-$ interactions produced by ^8B solar neutrinos at the Sudbury neutrino observatory. Phys. Rev. Lett. **87**, 071301 (2001)
17. S.P. Martin, A supersymmetry primer 1–98 (1997) [Adv. Ser. Dir. High Energy Phys. **18**, 1 (1998)]

© The Author(s), under exclusive license to Springer Nature Switzerland AG 2020
S. S. Ghosh et al., *General Model Independent Searches for Physics Beyond the Standard Model*, SpringerBriefs in Physics,
https://doi.org/10.1007/978-3-030-53783-8

18. A.H. Chamseddine, R. Arnowitt, P. Nath, Locally supersymmetric grand unification. Phys. Rev. Lett. **49**, 970–974 (1982)
19. R. Barbieri, S. Ferrara, C.A. Savoy, Gauge models with spontaneously broken local supersymmetry. Phys. Lett. B **119**(4), 343–347 (1982)
20. L. Hall, J. Lykken, S. Weinberg, Supergravity as the messenger of supersymmetry breaking. Phys. Rev. D **27**, 2359–2378 (1983)
21. M. Gell-Mann, P. Ramond, R. Slansky, Complex spinors and unified theories. Conf. Proc. **C790927**, 315–321 (1979)
22. T. Yanagida, Horizontal gauge symmetry and masses of neutrinos. Conf. Proc. **C7902131**, 95–99 (1979)
23. T. Kaluza, Zum Unitaritätsproblem der Physik. Sitzungsber. Preuss. Akad. Wiss. Berlin (Math. Phys.) **1921**, 966–972 (1921) [Int. J. Mod. Phys. D **27**(14), 1870001 (2018)]
24. O. Klein, Quantentheorie und fünfdimensionale Relativitätstheorie. Zeitschrift für Physik **37**(12), 895–906 (1926)
25. N. Arkani-Hamed, S. Dimopoulos, G.R. Dvali, The hierarchy problem and new dimensions at a millimeter. Phys. Lett. B **429**, 263–272 (1998)
26. I. Antoniadis, N. Arkani-Hamed, S. Dimopoulos, G.R. Dvali, New dimensions at a millimeter to a Fermi and superstrings at a TeV. Phys. Lett. B **436**, 257–263 (1998)
27. I. Antoniadis, S. Dimopoulos, G.R. Dvali, Millimeter range forces in superstring theories with weak scale compactification. Nucl. Phys. B **516**, 70–82 (1998)
28. A. Pérez-Lorenzana, An introduction to extra dimensions. J. Phys.: Conf. Ser. **18**, 224–269 (2005)
29. M. Perelstein, Little Higgs models and their phenomenology. Prog. Part. Nucl. Phys. **58**, 247–291 (2007)
30. R.D. Peccei, The strong CP problem and axions. Lect. Notes Phys. **741**, 3–17 (2008) [3 (2006)]
31. G. Alimonti et al., The Borexino detector at the Laboratori Nazionali del Gran Sasso. Nucl. Instrum. Methods **A600**, 568–593 (2009)
32. X. Guo et al., A precision measurement of the neutrino mixing angle θ_{13} using reactor antineutrinos at Daya-Bay (2007)
33. M. Apollonio et al., Limits on neutrino oscillations from the CHOOZ experiment. Phys. Lett. B **466**, 415–430 (1999)
34. K. Eguchi et al., First results from KamLAND: evidence for reactor anti-neutrino disappearance. Phys. Rev. Lett. **90**, 021802 (2003)
35. Y. Itow et al., The JHF-Kamioka neutrino project, in *Neutrino Oscillations and Their Origin. Proceedings, 3rd International Workshop, NOON 2001, Kashiwa, Tokyo, Japan, December 508, 2001* (2001), pp. 239–248
36. Y.-F. Li, Overview of the Jiangmen underground neutrino observatory (JUNO). Int. J. Mod. Phys. Conf. Ser. **31**, 1460300 (2014)
37. R. Acciarri et al., Long-baseline neutrino facility (LBNF) and deep underground neutrino experiment (DUNE) (2016)
38. M.G. Aartsen et al., The IceCube neutrino observatory: instrumentation and online systems. JINST **12**(03), P03012 (2017)
39. M. Ageron et al., ANTARES: the first undersea neutrino telescope. Nucl. Instrum. Methods **A656**, 11–38 (2011)
40. E. Aprile, K. Arisaka, F. Arneodo, A. Askin, L. Baudis, A. Behrens, E. Brown, J.M.R. Cardoso, B. Choi, D. Cline, S. Fattori, A.D. Ferella, K.L. Giboni, A. Kish, C.W. Lam, R.F. Lang, K.E. Lim, J.A.M. Lopes, T. Marrodán Undagoitia, Y. Mei, A.J. Melgarejo Fernandez, K. Ni, U. Oberlack, S.E.A. Orrigo, E. Pantic, G. Plante, A.C.C. Ribeiro, R. Santorelli, J.M.F. dos Santos, M. Schumann, P. Shagin, A. Teymourian, E. Tziaferi, H. Wang, M. Yamashita, The XENON100 dark matter experiment. Astropart. Phys. **35**(9), 573–590 (2012)
41. K.T. Lesko, The Sanford underground research facility at Homestake. Eur. Phys. J. Plus **127**(9), 107 (2012)

42. R. Bernabei, P. Belli, F. Cappella, R. Cerulli, C.J. Dai, A. d'Angelo, H.L. He, A. Incicchitti, H.H. Kuang, J.M. Ma, F. Montecchia, F. Nozzoli, D. Prosperi, X.D. Sheng, Z.P. Ye, First results from DAMA/LIBRA and the combined results with DAMA/NaI. Eur. Phys. J. C **56**(3), 333–355 (2008)

43. D.S. Akerib et al., Exclusion limits on the WIMP-nucleon cross section from the first run of the cryogenic dark matter search in the Soudan underground laboratory. Phys. Rev. D **72**, 052009 (2005)

44. Z. Ahmed, D.S. Akerib, S. Arrenberg, M.J. Attisha, C.N. Bailey, L. Baudis, D.A. Bauer, J. Beaty, P.L. Brink, T. Bruch, R. Bunker, S. Burke, B. Cabrera, D.O. Caldwell, J. Cooley, P. Cushman, F. DeJongh, M.R. Dragowsky, L. Duong, J. Emes, E. Figueroa-Feliciano, J. Filippini, M. Fritts, R.J. Gaitskell, S.R. Golwala, D.R. Grant, J. Hall, R. Hennings-Yeomans, S. Hertel, D. Holmgren, M.E. Huber, R. Mahapatra, V. Mandic, K.A. McCarthy, N. Mirabolfathi, H. Nelson, L. Novak, R.W. Ogburn, M. Pyle, X. Qiu, E. Ramberg, W. Rau, A. Reisetter, T. Saab, B. Sadoulet, J. Sander, R. Schmitt, R.W. Schnee, D.N. Seitz, B. Serfass, A. Sirois, K.M. Sundqvist, M. Tarka, A. Tomada, G. Wang, S. Yellin, J. Yoo, B.A. Young, Search for weakly interacting massive particles with the first five-tower data from the cryogenic dark matter search at the Soudan underground laboratory. Phys. Rev. Lett. **102**, 011301 (2009)

45. J. Aalbers et al., DARWIN: towards the ultimate dark matter detector. JCAP **1611**, 017 (2016)

46. K. Zioutas et al., First results from the CERN axion solar telescope (CAST). Phys. Rev. Lett. **94**, 121301 (2005)

47. S.J. Asztalos et al., A SQUID-based microwave cavity search for dark-matter axions. Phys. Rev. Lett. **104**, 041301 (2010)

48. P. Brun et al., A new experimental approach to probe QCD axion dark matter in the mass range above 40 μeV. Eur. Phys. J. C **79**(2), 186 (2019)

49. C.P. Salemi, First results from ABRACADABRA-10cm: a search for low-mass axion dark matter, in *54th Rencontres de Moriond on Electroweak Interactions and Unified Theories (Moriond EW 2019) La Thuile, Italy, 16–23 March 2019* (2019)

50. G. Arnison et al., Experimental observation of isolated large transverse energy electrons with associated missing energy at $\sqrt{s}$ = 540 GeV. Phys. Lett. B **122**, 103–116 (1983)

51. R. Brandelik et al., Evidence for planar events in e+ e- annihilation at high-energies. Phys. Lett. **86B**, 243–249 (1979)

52. C. Berger et al., Evidence for gluon Bremsstrahlung in e+ e- annihilations at high-energies. Phys. Lett. **86B**, 418–425 (1979)

53. D.P. Barber et al., Discovery of three jet events and a test of quantum chromodynamics at PETRA energies. Phys. Rev. Lett. **43**, 830 (1979)

54. W. Bartel et al., Observation of planar three jet events in e+ e- annihilation and evidence for gluon Bremsstrahlung. Phys. Lett. **91B**, 142–147 (1980)

55. D. Decamp et al., ALEPH: a detector for electron-positron annihilations at LEP. Nucl. Instrum. Methods **A294**, 121–178 (1990) [Erratum: Nucl. Instrum. Methods **A303**, 393 (1991)]

56. P.A. Aarnio et al., The DELPHI detector at LEP. Nucl. Instrum. Methods **A303**, 233–276 (1991)

57. B. Adeva et al., The construction of the L3 experiment. Nucl. Instrum. Methods **A289**, 35–102 (1990)

58. K. Ahmet et al., The OPAL detector at LEP. Nucl. Instrum. Methods **A305**, 275–319 (1991)

59. F. Abe et al., The CDF detector: an overview. Nucl. Instrum. Methods **A271**, 387–403 (1988)

60. S. Abachi et al., The D0 detector. Nucl. Instrum. Methods **A338**, 185–253 (1994)

61. V.M. Abazov et al., The upgraded D0 detector. Nucl. Instrum. Methods **A565**, 463–537 (2006)

62. F. Abe et al., Observation of top quark production in $\bar{p}p$ collisions. Phys. Rev. Lett. **74**, 2626–2631 (1995)

63. S. Abachi et al., Observation of the top quark. Phys. Rev. Lett. **74**, 2632–2637 (1995)

64. I. Abt et al., The H1 detector at HERA. Nucl. Instrum. Methods **A386**, 310–347 (1997)

65. B. Aubert et al., The BaBar detector. Nucl. Instrum. Methods **A479**, 1–116 (2002)

66. A. Abashian et al., The Belle detector. Nucl. Instrum. Methods Phys. Res. Sect. A: Accel. Spectrom. Detect. Assoc. Equip. **479**(1), 117–232 (2002). Detectors for asymmetric B-factories

67. ATLAS Collaboration, The ATLAS experiment at the CERN large hadron collider. J. Instrum.
    3(08), S08003–S08003 (2008)
68. C.M.S. Collaboration, The CMS experiment at the CERN LHC. J. Instrum. **3**(08), S08004–
    S08004 (2008)
69. LHCb Collaboration, The LHCb detector at the LHC. J. Instrum. **3**(08), S08005–S08005 (2008)
70. CMS Collaboration, MUSiC, a model unspecific search for new physics, in pp collisions at
    $\sqrt{s} = 13$ TeV (XXX) (2020)
71. B. Abbott et al., Search for new physics in $e\mu x$ data at dø using Sleuth: a quasi-model-
    independent search strategy for new physics. Phys. Rev. D **62**, 092004 (2000)
72. V.M. Abazov et al., Quasi-model-independent search for new physics at large transverse
    momentum. Phys. Rev. D **64**, 012004 (2001)
73. B. Abbott et al., Quasi-model-independent search for new high p_T physics at d0. Phys. Rev.
    Lett. **86**, 3712–3717 (2001)
74. V.M. Abazov et al., Model independent search for new phenomena in $p\bar{p}$ collisions at $\sqrt{s} = $
    1.96 TeV. Phys. Rev. D **85**, 092015 (2012)
75. T. Aaltonen et al., Model-independent and quasi-model-independent search for new physics at
    CDF. Phys. Rev. D **78**, 012002 (2008)
76. T. Aaltonen et al., Global search for new physics with 2.0 fb^{-1} at CDF. Phys. Rev. D **79**, 011101
    (2009)
77. A. Aktas et al., A general search for new phenomena in ep scattering at HERA. Phys. Lett. B
    602(1), 14–30 (2004)
78. F.D. Aaron et al., A general search for new phenomena at HERA. Phys. Lett. B **674**, 257–268
    (2009)
79. M. Aaboud et al., A strategy for a general search for new phenomena using data-derived signal
    regions and its application within the ATLAS experiment. Eur. Phys. J. C **79**(2), 120 (2019)
80. P. Langacker, The physics of heavy Z' gauge bosons. Rev. Mod. Phys. **81**, 1199–1228 (2009)
81. T. Hebbeker, A global comparison between L3 data and standard model Monte Carlo - a first
    attempt. L3 note 2305 (1998)
82. CMS Collaboration, MUSiC - an automated scan for deviations between data and Monte Carlo
    simulation. Technical report CMS-PAS-EXO-08-005, CERN, Geneva (2008)
83. CMS Collaboration, MUSiC, a model unspecific search for new physics, in pp collisions at
    $\sqrt{s} = 8$ TeV. Technical report CMS-PAS-EXO-14-016, CERN, Geneva (2017)
84. CMS Collaboration, MUSiC, a model unspecific search for new physics, in pp collisions
    at sqrt(s)=13 TeV. Technical report CMS-PAS-EXO-19-008, CERN, Geneva (2020). To be
    published
85. G. Altarelli, B. Mele, M. Ruiz-Altaba, Searching for new heavy vector bosons in $p\bar{p}$ colliders.
    Z. Phys. C **47**, 676 (1998)
86. T. Sjöstrand, S. Mrenna, P. Skands, PYTHIA 6.4 physics and manual. JHEP **05**, 026 (2006)
87. T. Sjöstrand, S. Mrenna, P. Skands, A brief introduction to PYTHIA 8.1. Comput. Phys. Com-
    mun. **178**, 852 (2008)
88. C.M.S. Collaboration, Event generator tunes obtained from underlying event and multiparton
    scattering measurements. Eur. Phys. J. C **76**, 155 (2016)
89. G. Choudalakis, On hypothesis testing, trials factor, hypertests and the BumpHunter, in *Pro-
    ceedings, PHYSTAT 2011 Workshop on Statistical Issues Related to Discovery Claims in Search
    Experiments and Unfolding, CERN, Geneva, Switzerland 17–20 January 2011* (2011)
90. J.H. Collins, K. Howe, B. Nachman, Anomaly detection for resonant new physics with machine
    learning. Phys. Rev. Lett. **121**(24), 241803 (2018)
91. A. Blance, M. Spannowsky, P. Waite, Adversarially-trained autoencoders for robust unsuper-
    vised new physics searches. JHEP **10**, 047 (2019)
92. ATLAS Collaboration, Dijet resonance search with weak supervision using $\sqrt{s} = 13$ TeV pp
    collisions in the ATLAS detector, arXiv:2005.02983